MENTAL ILLNESS: LAW AND PUBLIC POLICY

PHILOSOPHY AND MEDICINE

Editors:

H. TRISTRAM ENGELHARDT, JR.

Kennedy Institute of Ethics, Georgetown University, Washington, D.C., U.S.A.

STUART F. SPICKER

University of Connecticut Health Center, Farmington, Conn., U.S.A.

VOLUME 5

MENTAL ILLNESS:
LAW AND PUBLIC POLICY

Edited by

BARUCH A. BRODY

Dept. of Philosophy, Rice University, Houston, Texas, U.S.A.

and

H. TRISTRAM ENGELHARDT, JR.

Kennedy Institute of Ethics, Georgetown University,
Washington, D.C., U.S.A.

SPRINGER-SCIENCE+BUSINESS MEDIA, B.V.

Library of Congress Cataloging in Publication Data

CIP

Main entry under title:

Mental illness.

(Philosophy and medicine ; v. 5)

Developed from a series of six lectures sponsored by Rice University and the University of Texas Medical Branch at Galveston in the fall of 1976.

Includes bibliographies and index.

1. Insanity—Jurisprudence—Addresses, essays, lectures. 2. Mental health laws—Addresses, essays, lectures. 3. Insane—Commitment and detention—Addresses, essays, lectures. I. Brody, Baruch A. II. Engelhardt, Hugo Tristram, 1941– III. William Marsh Rice University, Houston, Tex. IV. University of Texas Medical Branch at Galveston.

[DNLM: 1. Mental disorders—United States—Legislation. 2. Public policy—United States. W3 PH609 v. 5 / WM33 AA1 M5]

K640.M4 344'.044 80–11715

ISBN 978-94-009-8974-0 ISBN 978-94-009-8972-6 (eBook)

DOI 10.1007/978-94-009-8972-6

EDITORIAL PREFACE

This volume developed from and around a series of six lectures sponsored by Rice University and the University of Texas Medical Branch at Galveston in the Fall of 1976. Though these lectures on the concepts of mental health, mental illness and personal responsibility, and the social treatment of the mentally ill were given to general audiences in Houston and Galveston, they were revised and expanded to produce six extensive formal essays by Dan Brock, Jules Coleman, Joseph Margolis, Michael Moore, Jerome Neu, and Rolf Sartorius. The five remaining contributions by Daniel Creson, Corinna Delkeskamp, Edmund Erde, James Speer, and Stephen Wear were in various ways engendered by the debates occasioned by the original six lectures. In fact, the majority of the last five contributions emerged from informal discussions occasioned by the original lecture series.

The result is an interlocking set of essays that address the law and public policy insofar as they bear on the treatment of the mentally ill, special attention being given to the definition of mental illness, generally and in the law, to the issues of the bearing of mental incompetence in cases of criminal and civil liability, and to the issue of involuntary commitment for the purpose of treatment or for institutional care. There is as well a critical defense of Thomas Szasz's radical proposal that mental illnesses are best understood as problems in living, not as diseases. This analysis by Corinna Delkeskamp, as well as the first six essays, places the legal issues within general philosophical and cultural contexts in order to offer a fundamental review and reappraisal of our views of mental illness and their bearing on law and public policy. As the closing four commentaries show in their review of the essays in the first part of this volume, these problems can only be understood if they are recognized as basic conceptual issues regarding the boundary lines between medicine and the law, between treatment and punishment, between freedom and determinism, and between health and illness. These perennial problems arise in the course of the incremental reinvention of the scope of various social institutions and practices, given reappraisals of various human goods and goals, such as liberty, toleration of deviance, concerns for the goods of others, as well as reassessments of such endeavors as the treatment of mental illness and the punishment of criminals.

B. A. Brody and H. Tristram Engelhardt, Jr. (eds.), Mental Illness: Law and Public Policy. v–vi.

As usual, basic questions unearth basic conceptual problems and riddles, the examination of which defines the history, if not the progress, of a culture. It is appropriate that this volume should emerge from a series of lectures supported by the Texas Committee for the Humanities and Public Policy, a state-based committee of the National Endowment for the Humanities. These essays offer basic inquiries into a cluster of major issues of public policy in the light of reflections from the humanities. We are grateful to Rice University and the Institute for the Medical Humanities of the University of Texas Medical Branch at Galveston for having hosted these lectures. This interdisciplinary cooperation between a university and a medical school set a context for discussion that could encompass the range of issues that the authors of these essays address. These issues are timely and of public concern, as reflected by the fact that the lectures from which these essays developed were originally prepared for and presented to a primarily non-university audience. The problems addressed constitute a special sub-set of quandaries in the philosophy of medicine, issues currently in great dispute: the boundary line between medicine and law, and the similarities and dissimilarities between physical and mental diseases.

A volume that develops through stages has many contributors who play special roles other than that of a principal contributor. This volume is the result of the labor of many persons who have generously and importantly assisted in its preparation as well as supporting the series of lectures that was its ancestor. We wish to thank William B. Bean, Jane Backlund, John P. Billings, Chester R. Burns, Susan M. Engelhardt, Nolen Holcomb, Gammon Jarrell, Judge Jerome Jones, Joel B. Kirkpatrick, Esq., Ronald T. Luke, Jo Monaghan, John Moskop, Ruth Walker Moskop, Ralph H. Shuffler, and Stuart Spicker. To those and to all who contributed we here express our appreciation. We wish as well to acknowledge our debt to the D. Reidel Publishing Company, which assisted in many ways, including supporting a part of the publicity costs involved in announcing the original lecture series.

June 3, 1979

BARUCH A. BRODY
H. TRISTRAM ENGELHARDT, JR.

TABLE OF CONTENTS

SECTION V / CRITICAL COMMENTARIES

H. TRISTRAM ENGELHARDT, JR.

INTRODUCTION

Medicine and the law are two major social institutions, each supporting various and often quite disparate practices. It is frequently unclear where certain practices fall – whether they are truly medical or really legal endeavors, whether they are attempts to cure or care for persons with diseases, or attempts to punish criminals and rectify harms. Indeed, the practices intertwine in endeavors that bridge these major social institutions, where legal concerns for rectifying harms and medical concerns for cure and care are joined, as in the case of much public policy bearing on the mentally ill. However, precisely at such junctures one has grounds for concern in that each institution may subvert the goods of the other and of itself. Thus the law may become an instrument of medicine gaining unconsented-to power to medicalize life and thus pervert the law's function of securing and preserving more general social goods and rights. On the other hand, medicine may employ the law to constrain individuals to be treated; in doing so it may render medicine simply an instrument of social policy.

These concerns are complicated in the area of mental health, where norms of health and disease are unclear and therefore open to intentional distortion as instruments to suppress social deviance. Medicine is, after all, performative in its use of diagnosis. That is, diagnoses not only describe reality, they create reality as a government creates ambassadors through naming them. Medicine places individuals in sick roles, relieving them of some responsibilities, while vesting them with new responsibilities and rights. A sick individual may be excused from work or from criminal responsibility, yet be obliged to seek treatment, while receiving special rights, for example, to social welfare payments in virtue of being disabled. One might think here of Talcott Parsons's description of the sick role, in which the sick individual is (1) relieved of certain responsibilities, (2) held not to be guilty for persisting in the incapacities of the illness, so long as he or she (3) seeks treatment from (4) those who are curers or carers for such problems [11, 12, 13, 14]. This view of the sick role does not preclude blaming individuals for becoming ill. Thus, one may excuse from work an individual who develops bronchogenic carcinoma of the lung after years of smoking five packs of unfiltered cigarettes a day, despite understanding the health hazards involved. One could recognize that the

B. A. Brody and H. Tristram Engelhardt, Jr. (eds.), Mental Illness: Law and Public Policy. ix–xvii.

individual was not immediately responsible for the incapacities of the illness; the incapacities are not willfully displayed incapacities, as in the case of a malingerer. One might even hold that the individual acted correctly in seeking expert medical advice and help as soon as the disease manifested itself. However, one could still blame the individual for the imprudent past history of smoking; one could blame the individual for becoming ill. The same line of argument could be applied in the case of an alcoholic or drug addict, even if one holds such states to be mental illnesses, not willful acts of misconduct. The same could hold for other mental illnesses if avoidable predisposing factors become known.

The social role of being mentally ill is, however, even more powerful than the sick role imposed by physical illness. Imposing the former usually leads to more dramatic social changes for the bearer than occurs with most sick roles acquired by physical illnesses. Compare, for example, the impact of sick roles such as being asthmatic, having heart disease, having the flu, versus being hysterical, being schizophrenic, and being neurotic. The physically based sick roles, which come close to having such power, are those of having cancer, being terminally ill, being syphilitic, and being a leper [5]. The power of these more forceful sick roles depends on a conjoined moral disapproval or an extreme non-moral negative evaluation of the state involved. Most persons are very fearful of terminal illness. Moreover, for various reasons, cancer is so feared that people hesitate to admit that they have cancer, though heart disease is received as a much more acceptable affliction. One sees this reflected in the fact that obituaries will more often acknowledge that an individual has died of heart disease than of cancer. In the case of cancer, a circumlocution is often used: "He died after a prolonged illness."

Undoubtedly, mental illness is so disvalued because it strikes at the very roots of our personhood. It visits us with uncontrollable fears, obsessions, compulsions, and anxieties. Mental illness gets inside of us. It is much more difficult to stand at a distance from one's neurotic anxieties than from one's athlete's foot or cold. Mental illness tends more intrusively and unavoidably to affect our view of ourselves and of our surrounding world. That is to say, the world of hysterics, neurotics, or schizophrenics is more clearly constituted by their illnesses than is the world of most of the physically ill. This is captured in part by the language we use in describing the mentally ill. One *is* an hysteric, *is* a neurotic, *is* an obsessive, *is* a schizophrenic, *is* a manic-depressive. On the other hand, one *has* heart disease, *has* cancer, *has* the flu, *has* malaria, *has* smallpox. The differences are surely matters of degree, not of kind. And there are surely many physical diseases that have the same sense of pervasive permeation of the patient's life. For example, one *is* a diabetic. *Being* a diabetic often commits one to an encompassing way of life, as does a great deal of mental illness. Also, one distinguishes those who are neurotic from those who simply have a character disorder. However, mental illnesses are quite often perceived, though not always, as especially invasive and constitutive of the very character of the person who has them. Thus diagnoses of mental illness have a major impact on law and public policy, because, rightly or wrongly, mental illness is seen to define its bearers as persons. If one cannot determine one's view of the good, but is rather driven by a compulsion, overwhelmed by uncontrollable fears, or so disordered in one's thought as not to be able to conceive of, much less to execute, an action as *one's* chosen action, then one fails to be a moral agent. One is not to blame for what one does; one is not the author of one's action. Hence the diagnosis of mental illness can be invoked in a court of law in an attempt to show that one is not culpable for one's actions, not worthy of blame or praise, not likely to be deterred by the threat of punishment, or not aided by a prison's reformative measures. But as Michael Moore, Jerome Neu, and Corinna Delkeskamp show, this special use of the sick role is an extremely precarious enterprise for law and medicine. That is, medicine becomes a potentially powerful instrument of social policy, open to special abuse, as the law draws upon notions of mental illness in order to fashion lines between legal culpability and legal blamelessness.

Concerns about the power of medicine are thus heightened in the case of mental health care, where, through the law, categories of mental illness can be used to remove individuals (sometimes in a very radical fashion) from the general society. Health care institutions for the mentally ill do, after all, offer an alternative holding-ground for those who violate societal norms. Thus, psychopathic murderers are not punished, they are rather placed in mental health care institutions. This appears to accord with a general social consensus that even if a person causes a great harm (e.g., a paranoid schizophrenic kills an innocent individual) for which that person is not to blame because he or she is no longer a moral agent, then one ought not to punish that person. In fact, in a strict sense one cannot punish that individual. He or she is not worthy of blame. There is no evil intent, *mens rea*, to blame. Nor, perhaps, are such individuals likely to be deterred or reformed by the usual institutions of punishment.

However, members of the public often wish at least to quarantine such persons in order to be secure from further harm. Even if the public could be certain that a paranoid schizophrenic murderer had been promptly cured, so that he or she posed no greater threat then the average normal individual, there would likely be many who would be distressed if such "murderers" were let loose. The reasons for such chagrin are likely to be multiple. One may suspect that the accused used the mental illness sick role to escape punishment while in fact being culpable, and that in addition his or her successful escape from justice will encourage others to try to do likewise. Or, perhaps, one might believe that many schizophrenics respond sufficiently to the reality principle so as to be at least in part deterred by the threat of punishment. Or perhaps one covertly views the institutionalization of the mentally ill as a special way of dealing with social deviants, a point that Daniel Creson examines in some detail. Or one might feel that the murdered person or his family deserve more vengeance on their behalf. This last view may also lie behind the practice of assigning tort liability for the mentally ill. If the mentally ill harm others, then the goods owned by the mentally ill injurer should be used (so such an intuition would dictate) to make the injured person whole, even if the injurer was in no sense 'guilty' of a culpable action (i.e., was not morally worthy of blame). In short, a function of the category 'mental illness' is to draw a line between institutions of care and cure, and institutions of punishment and recompense. This function is likely to be complex, for the rubric 'mental illness' can be used to serve multiple purposes for the law.

Because these purposes evolve over time and often without sufficient analytic care, they can come to allow unexpected abuses, fail to support the originally endorsed goals, overlook important goals, or assume irrelevant functions, depending upon unjustifiable distinctions. As a result, many of the essays in this volume attend to the interplay of medicine and law over time. Michael Moore, for example, traces the impact that Isaac Ray's *Treatise on the Medical Jurisprudence of Insanity* [15], as well as various views of the causal force of mental diseases, had upon the development of legal tests for insanity. Moore underscores the confusions that arose out of thinking that judgments of mental illness are simply descriptive, not also normative, that medicine's use of mental illness is the same as the law's, or that showing that a criminal action is a product of mental disease establishes that the criminal is not blameworthy. For the law the function of the rubric 'mentally ill' or 'insane' is thus ambiguous, for these points have been and are understood variously in different interpretations in the law. One is moved, then, as these essays suggest, to a reappraisal of the concept of mental illness and its functions for public policy.

Jules Coleman, for example, suggests that the legal concept of mental illness or insanity could in some circumstances be expanded to defeat the tort liability of the mentally ill. If a mentally ill individual is so impaired as not to be worthy of blame or praise, does not such a person then fail to be an actor, an agent? Should that individual not be viewed as relieved of fault for his or her actions? Should not their harmful actions be like the act of God in a strong wind, bringing down a sturdy tree upon a passerby? There is physical harm, but no fault, no liability. Thus, Coleman suggests that tort liability, with respect to the mentally ill, should be an appraisal not only of acts but of actors when making judgments of fault.

The status (or lack thereof) of the mentally ill as agents is core, as the discussions in this volume show, to many of the ways in which the law acts upon and for the mentally ill. If the mentally ill are not moral agents, then one cannot offend the principle of respect for freedom in imposing involuntary commitment or in acting paternalistically on their behalf. If one, for example, understood respect for persons as the necessary condition for the possibility of a community based not on force but on regard for the freedom of others, one need not respect the severely mentally ill as members of such a moral community. There is no freedom to be respected as one might be constrained to respect the bizarre and capricious decisions of a competent person. There are simply interests of that individual to protect, and therefore one may act on his best interests without his permission, for there is no permission to be had.

However, as Jerome Neu and others in this volume argue, freedom is a matter of degree. Such a position engenders further problems. How does one come to give an account of partial freedom or competence? Is it competence in some areas not in others, or is it a matter of degrees of competence? The first is an easier issue, for it allows one still to speak of the competent/incompetent dichotomy in a binary fashion. One either has it or not – the only question is with regard to what areas of one's life one is competent. However, the language of degrees of competence is more puzzling. What does it mean, for example, to be partially criminally culpable, though such issues are weighed, for example, in giving sentences for crimes.

Presuming some sense to the problem of partial competence, one may ask when and how far and to what extent may one act to achieve what one takes to be the best interests of an incompetent. When does the state become as a parent to a child, and enter to cure and secure care for an incompetent? Or when does the state under the guise of the kind parent use the rubric of mental illness as a vehicle for illicitly controlling social deviance? The contrariness of our intuitions on this issue is reflected in the fact that we are more likely to confine a paranoid schizophrenic who orally threatens the life of another, than we are to confine a competent individual who also simply orally threatens another, even if we were to know that the latter posed in actuality a much greater threat. The practices of committing the mentally ill for treatment, or because we suspect them to be a danger to others or to themselves, thus invites a careful reappraisal. These concerns have been reflected, for example, in a recent Supreme Court decision that requires the 'clear and convincing' standard of proof in a civil proceeding for involuntary commitment to a state mental hospital for an indefinite period [1].

It is far from clear the extent to which practices of commitment exist to protect the best interests of those who are not moral agents, to enforce prudent behavior on the foolish and deviant, or to quarantine the offensive and bizarre. One might in the case of offensive and bizarre behavior think of the long-term commitments often imposed on those who repeatedly expose themselves in public, a bothersome but not greatly harmful behavior. It is for such reasons that moral considerations concerning respect of freedom and utility of civil commitment mechanisms receive scrutiny in this volume by authors such as Rolf Sartorius, Dan Brock, Corinna Delkeskamp, Stephen Wear, and Edmund Erde. Because the public policy arguments are often framed with a lack of definition, so that it is not clear which considerations of respect of freedom or which measures of utility will lead to which practices of involuntary commitment, a number of the essays in this collection attempt to clarify the issue by analyzing different elements of such arguments. For example, Corinna Delkeskamp maintains that abolishing all involuntary civil commitment and holding the mentally ill fully responsible, a position explored by Thomas Szasz [16, 17], may be justified from utilitarian considerations as a way of signalling social problems that should be resolved prior to restricting the liberty of individuals. Towards this end Delkeskamp gives a reconstruction of elements of Szasz's position.

The practice of involuntary commitment for care or cure is problematic. One has on the one hand a dimension of medicine that for its clientele depends to a large degree upon force. Individuals are impressed into patienthood. On the other hand, mental health care offers, as Daniel Creson observes, an often-appealing alternative to formal criminal sanctions for aberrant behavior. Mental health care is thus, perhaps ineluctably, an extension of the modes of social control of deviance. There are undoubtedly limits to the amount of deviant behavior that society or relatives are likely to bear.

These issues are not amenable to easy resolution, though there may be good reasons to wish there were. For example, one might wish simply to treat mental illness as a myth, as Thomas Szasz argues [16, 17]. Still, as Stephen Wear observes, there appear to be many unquestionable cases of mental illness. There appear to be identifiable patterns of mental derangement so marked by their erosion of rationality that they appear in many ways similar to those derangements incident to somatic diseases (i.e., febrile delirium) or to intoxication. They appear to be constellations of signs and symptoms that are simply caused, not ways of life or problems of living freely assumed. Yet most categories of mental illness have successfully resisted generally acknowledged, causal explanations that would render such syndromes into well-established disease categories. However, syndromes have served somatic medicine in the absence of clearly established causal accounts; they have provided bases for prognosis, care, and at times even cure. One might simply notice that quinine helps malaria, even without a correct causal account of malaria. Still, mental disease categories seem so much more in disarray than somatic categories, that one is tempted to see the discrepancy between them and those of physical illness as a difference in kind, not just degree.

It is because of such considerations that disputes over the status of disease concepts are relevant here. It is with them that this volume opens. The recent philosophical literature has provided a series of reexaminations of the concept of illness and disease generally, not just mental illness and disease [2, 4, 6, 7, 8, 9, 10, 19]. These have explored the extent to which concepts of disease are not only discovered but also invented, the extent to which they depend upon evaluation and are not simply descriptive or explanatory in a value-free fashion. Insofar as concepts of disease reflect views of proper human functions, concepts of mental illness and disease will be open to revision as one alters one's views regarding proper human functions. In fact, considerations regarding the impact of particular views of health and disease are likely to suggest the advisability of reassessing some tacitly accepted views. After all, it was in part out of considerations of utility that Thomas Trotter originally argued that alcoholism be considered a disease [18]. The reexamination of the status of homosexuality as a disease shows as well the extent to which mental disease rubrics reflect social and political considerations. At the very least one should be chastened by the realization of the partially created nature of nosologies in psychiatry, not to mention the rest of medicine. The arguments that have arisen in the course of analyzing clinical judgment in medicine have shown the partially instrumental character of diagnosis [3]. Thus, if concepts of disease are best taken not as true or false but as more or less useful, as Henrik Wulff suggests [20], one might then come to reappraise how and to what purpose different psychiatric nosologies are useful. This volume is presented as a step in that direction.

These essays offer a reappraisal of the concept of mental illness as it supports public policies, especially those of civil commitment and of the criminal and civil liability of the mentally ill. As James Speer argues, such a reappraisal involves a fundamental reassessment of how the law should function in these areas – an extensive goal to which the present volume can only make a modest contribution. However, such reassessments are necessary if the institutions of law and medicine are to serve adequately the goals we humans choose through law and medicine. Since these goals develop through history, they commit us to an ongoing dialectic of analysis, criticism, and development. This volume, in integrating concepts from the philosophy of medicine and the law, and drawing from contributors educated in the law, philosophy, medicine, psychiatry, history, and anthropology, is intended as a contribution to our reflection on these goals and the bearing they have on present public policies toward the mentally ill.

June 3, 1979

BIBLIOGRAPHY

1. Addington v. State of Texas, 47 U.S.L.W. 4473 (1979).
2. Boorse, C.: 1975, 'On the Distinction Between Disease and Illness', *Philosophy and Public Affairs* 5, 49–68.
3. Engelhardt, H. T., Jr. and Spicker, S. F. (eds.): 1979, *Clinical Judgment: A Critical Appraisal*, D. Reidel Publ. Co., Dordrecht, Holland.
4. Engelhardt, H. T., Jr.: 1975, 'The Concepts of Health and Disease', in H. T. Engelhardt, Jr. and S. F. Spicker (eds.), *Evaluation and Explanation in the Biomedical Sciences*, D. Reidel Publ. Co., Dordrecht, Holland, pp. 125–141.
5. Foucault, M.: 1973, *Madness and Civilization*, Vintage Books, New York.
6. Grene, M.: 1977, 'Philosophy of Medicine: Prolegomena to the Philosophy', in P. Asquith and F. Suppe (eds.), *PSA 1976*, Philosophy of Science Association, East Lansing, Michigan, pp. 77–93.
7. King, L. S.: 1954, 'What is Disease?', *Philosophy of Science* 21, 193–203.
8. Kopelman, L.: 1975, 'On Disease: Theories of Disease and the Ascription of Disease', in H. T. Engelhardt, Jr. and S. F. Spicker (eds.), *Evaluation and Explanation in the Biomedical Sciences*, D. Reidel Publ. Co., Dordrecht, Holland, pp. 143–150.
9. Macklin, R.: 1972, 'Mental Health and Mental Illness: Some Problems of Definition and Concept Formation', *Philosophy of Science* 39, 341–365.
10. Margolis, J.: 1976, 'The Concept of Disease', *The Journal of Medicine and Philosophy* 1, 238–225.

11. Parsons, T.: 1958, 'Definitions of Health and Illness in the Light of American Values and Social Structure', in E. Jaco (ed.), *Patients, Physicians and Illness*, Free Press, Glencoe, Illinois, pp. 165–187.
12. Parsons, T.: 1958, 'Illness, Therapy and the Modern Urban American Family', in E. Jaco (ed.), *Patients, Physicians and Illness*, Free Press, Glencoe, Illinois, pp. 234–245.
13. Parsons, T.: 1957, 'The Mental Hospital as a Type of Organization', in M. Greenblatt, D. Levinson, and R. Williams (eds.), *The Patient and the Mental Hospital*, Free Press, Glencoe, Illinois, pp. 108–129.
14. Parsons, T.: 1951, *The Social System*, Free Press, New York, esp. pp. 248–279.
15. Ray, I.: 1962, *A Treatise on the Medical Jurisprudence of Insanity*, W. Overholser (ed.), Belknap Press, Cambridge, Massachusetts.
16. Szasz, T.: 1978, 'The Concepts of Mental Illness: Explanation or Justification?' in H. T. Engelhardt, Jr. and S. F. Spicker (eds.), *Mental Health: Philosophical Perspectives*, D. Reidel Publ. Co., Dordrecht, Holland, pp. 235–250.
17. Szasz, T.: 1961, *The Myth of Mental Illness*, Hoeber-Harper, New York.
18. Trotter, T.: 1804, *An Essay, Medical, Philosophical and Clinical, on Drunkenness and Its Effects on the Human Body*, 2nd ed., Longman, Hurst, Rees and Orine, London.
19. Wartofsky, M. W.: 1975, 'Organs, Organisms and Disease: Human Ontology and Medical Practice', in H. T. Engelhardt, Jr. and S. F. Spicker (eds.), *Evaluation and Explanation in the Biomedical Sciences*, D. Reidel Publ. Co., Dordrecht, Holland, pp. 67–83.
20. Wulff, H.: 1976, *Rational Diagnosis and Treatment*, Blackwell Scientific, Oxford.

SECTION I

THE CONCEPT OF MENTAL ILLNESS AND THE LAW

JOSEPH MARGOLIS

THE CONCEPT OF MENTAL ILLNESS: A PHILOSOPHICAL EXAMINATION

It is startling to reflect on a question of such obvious importance as the nature of mental illness, only to discover – as one inevitably does – that the most elementary or fundamental distinctions of relevance are still being radically disputed by competent professionals. Certainly, since the advent of Thomas S. Szasz's, *The Myth of Mental Illness* [32], the issue of whether the concept of mental illness is even coherent, let alone valid, has puzzled the entire therapeutic and health-care community as well as philosophers and social commentators, who, fair parasites that they are, sniff a good fight. Historically, of course, beginning with the work of Freud and his immediate precursors, notions of mental illness and mental disease were rather ambiguously linked with the undertaking of the medical profession. On the one hand, Freud did not regard psychoanalysis as a medical specialty; and on the other, it was, precisely, a group of professional physicians who first extended their practice to include varieties of psychoanalytic psychotherapy and various forms of proto-psychiatry. The trouble with deciding the medical legitimacy of the concept of mental illness – *a fortiori*, the legitimacy of extending medical care to the treatment of such illness – has always been the same, namely, that there are too many analogies and disanalogies between somatic medicine and the variety of disciplines that have centered on alleged mental disorders to decide the issue trimly. Freud himself always favored the possibility of a materialist reduction of his new theories, though increasingly, on review, he found the prospect ever slimmer; and those who have followed him have tended – usually without the same philosophical zeal – to adopt a sort of pragmatic dualism of psychic and physical forces.[1]

The usual objections against the concept of mental illness have been double-edged: the claim is made for instance that there is something inherently invalid about the notion itself; also, it is said that fostering the thesis that so-called mental illnesses may be medically treated is socially pernicious. The two themes appear conjoined in Szasz and are taken by him to be logically connected [32]. But it is of course entirely possible to view psychodynamically-oriented psychiatry and Freudian-oriented non-medical therapy as conceptually coherent but professionally undesirable, say, on the

B. A. Brody and H. Tristram Engelhardt, Jr. (eds.), Mental Illness: Law and Public Policy. 3–23.

grounds of their therapeutic ineffectiveness – broadly speaking, the argument of the behavior therapists [35]. There is, in fact, very good reason for tolerating both a medical and a non-medical interpretation of therapy directed to what, in a rather neutral sense, is called 'mental illness'. It should be noted that so-called mental illnesses span an enormous range of syndromes that, in relevant respects, are variably similar or dissimilar to standard physical illnesses and disorders: some, involving obvious incapacitation, as in what Freud himself regarded as hysterical paralysis, bear a striking resemblance to forms of physical incapacitation; others, like homosexuality, seem often to be judged as incapacity by some figurative reading of the prevailing norms and values of social existence [12], [1]. Short of a demonstration that there is indeed something intrinsically incoherent in the concept, the social implications of providing care for the incapacitated argue the need for flexibility in subsuming psychotherapy within the medical model – at the same time that one understands that the maneuver favoring medical inclusion or exclusion or both, for different 'disorders' within the range usually considered, is a practical matter concerned with mobilizing social resources for the care of those in genuine need.

There is good reason to think that it cannot be shown that psychotherapy is or is not, flatly, a medical discipline. The conceptual issues bearing on the plausible extension of terms under the pressure of changing classificatory purposes, including, not negligibly, a favorable public organization of putatively beneficial services, simply do not provide for the bare discovery of a false claim – which is not to say that they do not provide for dialectically weighted arguments about the good sense of thus including or excluding therapy, in whole or in part. A good deal of the argument against psychotherapy is based on a mistaken notion of the very nature of physical medicine. Were it clear that the concepts of health and disease restricted to somatic conditions are themselves inevitably controlled – however conservatively over the centuries – by a set of slowly changing social values [24], the inclusion or exclusion of psychotherapy would be seen at once as distinctly less significant conceptually than it has been supposed to be, both by its detractors and its champions.

The basic errors involved in precluding a medical construction of psychiatry and psychotherapy are most clearly evidenced by Szasz himself, despite all the nonsense that he and his associates have unquestionably disclosed. These errors may be tabulated quite straightforwardly.

(i) It is wrongly supposed that those who are said to be mentally ill are always and invariably shamming, assuming the role of the ill for some imagined advantage, and never actually incapacitated or unable to discard the seemingly disadvantaged role.

Szasz holds for example that mental illness is a 'counterfeit', that "*the similarity in appearance* [to genuine illness is] *deliberately created by a human operator* for some purpose" ([32], p. 39). But this is an empirical matter. Hence, apart from the evidence that some are actually victimized by their condition, even if they have induced it in themselves, it would be quite impossible to show, on the strength of this consideration, that psychiatry and psychotherapy were, construed as medical disciplines, conceptually incoherent or indefensible.

(ii) It is wrongly supposed that the difference between mental and physical illness lies at the very least in the fact that 'malfunctioning' of the first sort is simply social maladaptation – hence, judged so in terms of prevailing social norms and values; whereas malfunctioning of the second sort concerns the actual imperfect functioning of the human body without regard to social adaptation and social values – however it may produce secondary social malfunctioning.

Thus, Szasz holds that there are "two classes of disabilities, of 'problems' in meeting life. One category consists of bodily diseases – say, leprosy, tuberculosis, or cancer – which, by rendering imperfect the functioning of the human body as a machine, produce difficulties in social adaptation . . . the second category is characterized by difficulties in social adaptation not attributable to malfunctioning machinery but 'caused' rather by the purpose the machine was made to serve by those who built it (e.g., parents, society) or by those who use it, i.e., individuals" ([32], pp. 41–42). There are a fair number of challengeable claims secreted in these remarks. For one thing, Szasz appears to believe that psychogenic and sociogenic disorders, despite the fact that they may involve, in his own phrasing, 'malfunctioning machinery', malfunctions in the body, are rightly precluded from the category of illness or disease. For another, Szasz appears to believe that self-induced or socially induced maladaptation cannot, in principle, be suitably incapacitating to count as illness or disease. And finally, Szasz appears to believe that the categories of physical illness and disease are value-free and indifferent to the social values in terms of which social maladaptation is rightly so characterized.

Some of the countermoves to Szasz's implicit claims are obvious enough. Admitting a loose distinction between mental and physical illness corresponding to general medical practice but resisting, at this point, a full ontological account of the relation between the mental and the physical, it is apparent that so-called mental illnesses can be every bit as incapacitating as so-called physical disabilities; in fact, it is quite conceivable, as in organically based and (possibly) non-organically based schizophrenia, that very similar behavioral and conceptual syndromes obtain that are normally taken to be medically incapacitating. Aetiology is neither always clear nor always decisive. But certainly, in the absence of a valid and comprehensive account of the causal nature of disease factors – particularly with regard to elements provisionally sorted as mental and physical – it is pointless as well as hopeless to preclude psychogenic and sociogenic factors. The question of the value-free status of physical medicine deserves a separate review – to which we shall turn in a moment.

(iii) It is wrongly supposed that the classification of diseases is as such a matter of discovery rather than of useful invention and, in particular, that it is a matter of discovery on principles that conceptually preclude the admission of mental illness.

Szasz, as has already been noted, believes that mental illnesses are often 'very good . . . imitations' of physical illness ([32], pp. 39–40). But beyond this, on more clearly conceptual grounds, he distinguishes between the *discovery* of disease and the merely *ad hoc* invention or advocacy of construing certain patterns of social maladaptation as forms of mental illness. Thus he says: "I maintain that Freud did not 'discover' that hysteria was a mental illness. Rather, he advocated that so-called hysterics be declared 'ill'. The adjectives 'mental', 'emotional', and 'neurotic' are simply devices to codify – and at the same time obscure – the differences between the two classes of disabilities . . ." ([32], pp. 41–42). There is no doubt that Freud did *not* discover that hysteria was a form of illness. He did, indeed, advocate a change in the classificatory conception of disease in order to subsume a range of so-called mental illnesses within the medical model. But there is absolutely nothing conceptually suspicious about the maneuver. The thesis that classificatory divisions can somehow be 'discovered' as if they were the 'natural cuts' of the world has been disputed since Plato, and effectively dismissed as indemonstrable and questionable. All classificatory systems are schemes of convenience judged valid or at least reasonable in virtue of the explanatory or other related purposes for which given classificatory alternatives are introduced.[2]

In the case of Freud's innovation, two factors are of decisive importance. First of all, the admission of effective psychogenic and sociogenic forces – ironically, conceded by Szasz and other critics – actually accommodates whatever function medical classification may be assigned relative to the question of causal explanation. So, in the very spirit in which classification is often altered, for instance the classification of biological organisms relative to the genetics of populations and of evolution itself, the inclusion of mental illnesses within the medical model may, so far forth, be viewed as an effort to adjust nosology to a larger and more comprehensive range of forms of pertinent incapacitation. Secondly, to say that much, obliges us to consider that medicine is an art of ministering to humans and other organisms in terms of *pertinent* incapacitation; but the question of what constitutes pertinent incapacitation cannot possibly be answered solely or primarily in causal terms. What is wanted is a concept of the normal or characteristic or favored forms of human life with respect to which incapacitation, due to a variety of factors that could conceivably include, if we wished, the psychogenic and the sociogenic as aetiologically significant, may be judged as entitled to medically relevant exploration.

Here, then, we see that the resolution of the classificatory issue bears directly on the resolution of the question of the value-free or value-laden nature of both mental and physical illness. It might, in fact, be argued that the admission that medicine is a healing art entails that the concepts of mental and physical illness (and of health) are value-laden notions, unless it could be shown that the putative incapacitation to which the physician ministers is itself to be marked in an entirely value-free respect. It is extremely interesting to remember here that the Freudians, for their part, were quite convinced that the psychosexual functions to which they pointed and in terms of which the psychiatric nomenclature was developed were as value-neutral as the somatic functions in accord with which the classification of physical medicine has come to be systematized [8]. Furthermore, it is even more interesting to consider that the very reasons that betray the normative assumptions on which the entire venture of psychiatry depends – most notoriously, the Freudian contribution[3] – equally well betray the normative assumptions of physical medicine.

The issue, however, is a complex one and calls for the most careful analysis. Three very large but preliminary distinctions may be laid down as a resource toward an economical resolution. The first concerns the distinction between facts and values.[4] It is very often supposed, but erroneously, that factual judgments and value judgments differ in logical or grammatical form and, as a result, that it is impossible that one and the same judgment be both a value judgment and a factual judgment. But the simple truth is that a factual judgment is so called in virtue of its being possible to ascribe truth values to it, that is, the values 'true' and 'false' – or others like 'probably true' and 'probably false'. And a value judgment is so called in virtue of its predicates being valuational or normatively informed. For example, *if* the meaning of the predicates 'diseased', 'neurotic', 'intelligent', 'correct' are to be explained in terms of a theory of some normative condition or state, then the judgment employing those predicates is a value judgment. It is clear, therefore, that the distinction between factual and value judgments rests on a mixed classification and that it is entirely possible for the features of both to be found in a single judgment. Such a judgment may be termed a *finding:* medical, legal, moral, and aesthetic judgments provide us with a large number of characteristic findings.

But the distinction drawn is a purely formal one. It fails, consequently, to solve for us two further puzzles that bear directly on the answer to our original classificatory question: (a) whether or not the proper norms of medicine, law, morality, and the fine arts can, in some sense, be discovered by a direct inspection of the domain in question; and (b) whether the forms of incapacitation involved in the physician's art are conditions that can only be explained in terms of some suitable normative theory. Even so, if the foregoing argument sketch be admitted, then *it is quite irrelevant to the claim that the categories of medicine as such are value-laden or normatively freighted to point out that they are based on the specification of the natural functions of the human organism.* The non-sequitur is plain enough: on many views of mental and physical health, it is just the *normative functioning* of the human organism that is said in fact to be discovered.[5] So if, in Szasz's sense, physical medicine is concerned with the 'malfunctioning machinery' of the body, it does not in the least follow that the categories of physical medicine are value-free. The best that could be said, on the strength of this consideration, is that if physical medicine is normatively freighted, then the norms of the body's functions are straightforwardly open to cognitive discovery. But if that is the case, then, at the very least, the intended contrast between the concepts of mental and physical illness fails, unless it can be shown that, in fact, the norms of physical medicine *can* be discovered and those of psychiatry cannot be. In all honesty, however, this conjunction of propositions seems never to have actually been established. That it cannot be established is closely linked with the second of the preliminary distinctions promised.

The second distinction, then, concerns the fact that the human organism – the human being, if you will (though that expression is a source of equivocation, as we shall see) – can be classified as such, in any biologically relevant sense, without regard to the characteristic interests and personal or social values that different human communities may favor. It is, in short, not necessary, in classifying any creature as a member of *Homo sapiens*, to invoke the notion of excellence or merit or worth of any sort *qua* human being. All that is required is resemblance, on comparative grounds, to admitted physical specimens not themselves marked as excellent in any respect.[6] Hence, the idea that medicine, whether concerned with physical or mental illness and disease, can claim to *discover* the natural norms or normative functions of human beings clearly depends on a premise that has yet to be supplied, viz. concerning the connection between the biological classification of man and the discovery of his natural or normal mode of functioning. Strategically, then, two maneuvers must be considered: (1) how to show that the natural functions favored by physical medicine are not value-laden or normatively informed; (2) how to show, assuming such functions to be normatively informed, that they are nevertheless open to scientific discovery. Thus focused, criticism of the concept of mental health and illness favors the counterclaim (1′) that, as Szasz holds, social values are smuggled into the putative practice of psychiatric medicine – but not, in the same sense, in physical medicine; or (2′) that there are no natural 'mental functions' corresponding to determinate physical functions, that can fairly be claimed to be discovered.

Our third preliminary distinction follows on the heels of the second. There is no evidence of any distinctive faculty concerned with discerning values (as by way of analogy with sensory perception); and there is no evidence that the mere exercise of our perceptual faculties can, in a non-questionbegging way, be said to discern the natural values of things. The contrary doctrine has been termed cognitivism and may be dismissed as untenable [18]. But if so, then the provision of normative values of any sort – medical, legal, moral, aesthetic – are either frankly posited in accord with competing ideologies, doctrines, persuasions, ideals that are not themselves objectively confirmed or else are dialectically isolated in some way from among such alternative persuasions. The extent to which such norms can then be said to be objectively confirmed remains an uncertain matter. In the face of the constraints imposed by our three preliminary distinctions, it looks very much as if the wiser strategy would, if viable, be to claim that the functional considerations favored by physical medicine – and perhaps or perhaps not by psychotherapy and the relevant portions of psychiatry – are not value-laden at all.

One of the most sustained recent attempts to pursue this latter strategy has been made by Christopher Boorse [3]. In an earlier paper, on which he builds, Boorse had explicitly held that "With few exceptions, clinicians and philosophers are agreed that health is an essentially evaluative notion.

According to this consensus view, a value-free science of health is impossible. This thesis I believe to be entirely mistaken. I shall argue . . . that it rests on a confusion between the theoretical and the practical senses of 'health', or in other words, between disease and illness" ([2], p. 49). Boorse's account may serve as a useful stalking-horse.

Boorse distinguishes illness in the following way: "A disease is an *illness* only if it is serious enough to be incapacitating and therefore is

(i) undesirable for its bearer;
(ii) a title to special treatment; and
(iii) a valid excuse for normally criticizable behavior."[7]

(ii) and (iii) are relatively uncontroversial, since they signify practical consequences that follow from one's being ill, whatever illness may be and whatever disease may be – where disease may also justify (ii) and (iii). There are, it should be noted, diseases that are clinically perceptible that patients fail to distinguish, failing to note symptoms or failing to infer incapacitation or the threat of incapacitation. Alvan Feinstein, who terms such diseases lanthanic, is inclined to construe illness as, roughly, reflexively manifest disease, that is, disease discriminated in however attenuated a sense through sensations, proprioception, and the like [5]. This suggests a number of quibbles that one might entertain regarding Boorse's conception. For one thing, it is not altogether impossible to speak of illness (as of a certain feeling of incapacitation – which might then, to be sure, be incapacitating) without a state of disease's obtaining independently. If such feelings of illness were construed as sufficiently incapacitating to be a disease, then a considerable range of mental states that have been called into question by Szasz and others [30] would be entitled to another inning. Secondly, on Feinstein's view, it would be quite possible for a disease to be incapacitating and, being lanthanic, not perceived as such by a patient and hence not perceived as 'undesirable' by the patient; although, *qua* disease and *qua* incapacitating, it might well be undesirable. In that sense, the patient would be diseased but not ill. What this shows is that Boorse's condition (i) is somewhat unclear. (The counterpart unclarity in Feinstein concerns what would count as the relevant bounds of 'reflexively manifest disease'.) If (i) signifies 'undesirability' merely in the sense of actual or threatened incapacitation, then (i) might be no more than an analytic consequence of the agent's being in some diseased state; on the other hand, if (i) signifies 'undesirability' in the sense of a patient's recognizing to some extent his own diseased condition or of responding reflexively to some extent to the symptoms of his disease, then it is quite possible to be incapacitated in being diseased and yet not ill. But being incapacitated in some respect (for instance, being deliberately sterilized) may not be either undesirable (in whatever sense 'undesirable' may be understood) or perceived to be undesirable or a diseased state. So *if* a certain state is a disease, then it is (to that extent) undesirable though not necessarily incapacitating, and if it is incapacitating *qua* disease, then it is (on Boorse's view) undesirable. 'Undesirable' is a term with a curious history (harking back to Mill's *Utilitarianism*), since it has always been seen as encouraging an equivocation between a psychological sense and a normative sense: what is judged to be undesirable may not be reflexively perceived as such; and what is reflexively judged to be undesirable may, on some theory, not be undesirable at all. Boorse does not favor the psychological sense, and it appears that he does not favor the normative sense either. Hence, there is an initial puzzle about the sense in which to be incapacitated by disease is to be in an undesirable state; not that it is *not* undesirable, only that the sense of 'undesirable' is unclear. The point of these considerations is simply to indicate that Boorse's conception of illness is parasitic on his conception of disease and that the only condition that might have been substantive so far, namely, (i) is as yet of no independent help.

In fact, we may add several other counter-considerations of greater importance. One is this. It seems entirely possible that one may be ill relative to a diseased state, without being incapacitated; or alternatively, it is not in the least clear what would count as incapacitation – or, even what would count as sufficiently related to incapacitation to be judged a disease. It is certainly clear that one may be in a diseased state, say, in a state of incipient cancer which has not progressed very far: it is the probable threat of developing incapacitation causally linked to the organism's state that marks a disease, not merely actual incapacitation. But if this is so, then it is entirely possible – on Feinstein's model though not on Boorse's – to be ill but not actually incapacitated. It seems entirely fair to speak of degrees of one's state of disease (without requiring a measure of knowledge on the patient's part). Again, condition (i) seems to make sense only for sentient creatures or only for creatures of some rather high level of intelligence. But disease and incapacitation occur among plants and lower-order animals; yet we would not normally speak of illness among such organisms. Neither would Boorse, though if 'undesirable' is not used psychologically, it is not entirely clear why. Finally, but only provisonally (since we have not yet clarified Boorse's conception of disease), there is no doubt that the state of imminent death is characteristically incapacitating in the greatest degree; and yet, it is entirely possible that a person who is dying should be neither diseased nor ill. It is also reasonably clear that aging and dying may be judged and reflexively perceived to be undesirable because they are incapacitating. Hence, the burden of the entire argument rests on the concept of disease and of incapacitation by reason of disease. These quibbles are essentially designed to call into relief the ulterior concept of disease and the prospects of the concepts of mental illness.

This is what Boorse understands by health and disease, on which he intends to base his account of mental health, mental disease, and mental illness:

An organism is *healthy* at any moment in proportion as it is not diseased; and a *disease* is a type of internal state of the organism which:
(i) interferes with the performance of some natural function – i.e., some species-typical contribution to survival and reproduction – characteristic of the organism's age; and
(ii) is not simply in the nature of the species, i.e., is either atypical of the species or, if typical, mainly due to environmental causes ([3], pp. 62–63).

There are both defects and infelicities in Boorse's formulation, which are now no longer negligible since the fate of mental health and mental illness hangs in the balance.

For one thing, if we distinguish disease as an internal state normally expected to lead, for causal reasons, to incapacitation – a state that need not actually interfere with any function – we clearly suppose that organisms are diseased on grounds much weaker than Boorse's condition (i). Secondly, it is most certainly false or vacuous that diseases are so designated only if the putative interference is linked to some "species-typical contribution to survival and reproduction"; a great many diseases have nothing to do with survival or reproduction, except in the Pickwickian sense in which, should the disease prove fatal, it would indeed interfere with survival and reproduction. It is difficult to see that this qualification can be expected to distinguish between what Szasz calls 'malfunctioning machinery' and 'difficulties in social adaptation'. Consider, for example, that a mild cold is normally construed as a disease and is normally recognized in the manner of an illness; but it is entirely possible that the infection should be so mild that, even unattended medically, a 'normally healthy' person (now a puzzling notion) would shake it off before any interference or incapacitation with respect to survival or reproduction would occur. If, on the other hand, one attenuates sufficiently the connection between condition (i) and actual survival and reproduction, then it obviously becomes quite impossible to disqualify any internal state as a candidate disease state, solely on condition (i). What Boorse seems to intend here is a change of state affecting a process that makes a standard contribution to survival and reproduction rather than an actual interference with that process's contribution to survival and reproduction. If he does mean the former, then condition (i) is too strong, since a cold may temporarily affect processes contributing to survival and reproduction without interfering with their function in that regard; and if he means the latter, then condition (i) will preclude a mild cold from being either a disease or an illness, which seems arbitrary. But then, freckles affect processes contributing (in some sense) to survival and reproduction, without (usually) interfering with such functioning. In a great many cases, the 'contribution' to survival and reproduction affected will be 'species-typical', in whatever fair sense may be intended; but the change of state need not actually 'interfere' with that contribution. Nevertheless, sometimes the changed condition will be construed as a disease, sometimes not – not, however, in any way that is decisively and pertinently linked to conditions (i) and (ii). If, furthermore, we mean to use the model of somatic medicine to clarify the functions of the mind bearing on reproduction and survival, difficulties are bound to multiply. Would homosexuality for instance have to be viewed as a mental illness if heterosexual relations proved repugnant?

No sooner do we acknowledge this, than it becomes almost too plain that both survival and reproduction provide potentially equivocal criteria (though Boorse's intent is plain enough), for one must ask whether disease is to be specified relative to the survival and reproductive ability of the individual organism or to that of the stock. Neither alternative will be exclusively satisfactory, though Boorse inclines toward the first alternative [24]. For example, the incapacity to reproduce need not be judged to be a disease or illness in an individual since that incapacity need not interfere with any of the usual purposive activities (on any generous or restricted view) that an individual might be expected to engage in. It need not interfere with sexual activity and it need not interfere with 'parenting', that is, behaving as a parent; and it is preposterous to suppose (or at any rate obviously doctrinaire and hardly confirmable) that it is undesirable *tout court* that a given member of the human species not be able to reproduce or not be able to reproduce if, statistically, *most* of the members of that individual's age and sex group are able to reproduce. It is certainly not necessary as far as the race is concerned; and it seems very implausible to restrict our (medical) attention to the capacities of individuals with regard to survival and reproduction in such a way that whatever is to be regarded as a disease in individuals is somehow directly linked to those and only those two factors. It is difficult to consider survival and reproduction solely in individual terms, and it is even harder to see what the justification could be for characterizing as states of disease and illness temporally contingent deviations from some putative species-typical contribution to individual reproductive ability – 'undesirable' deviations regardless of a society's reflections on its reproductive needs. Also, it must be remembered that the population of the world is growing at an alarming rate; obviously, survival and reproduction as far as the race is concerned cannot be a deciding factor. (Boorse would not hold it to be such.) On the other hand, if one restricts population survival to the capacities of *particular political communities* (say, in terms of stable population size or favorable growth or even decline to a manageable level), then it is difficult to see how the matter of disease could be confined in purely biological or biologically functional terms. Some ideological intrusion is bound to occur. Also, if the survival and reproduction of a *species* require a *distinctive* genetic contribution on the part of sub-populations *within* a species, then it cannot be the case that what is essential for health must be "uniform within the species", as Boorse claims.[8] It is quite conceivable that the 'health' of a population may be favored by the sterility of a significant subpopulation *and* that our concept of the health of individuals be adjusted accordingly. On what grounds, then, may we say with assurance that involuntary sterility must be a form of disease, without regard to the conditions of survival of an entire stock? It seems very difficult to separate questions of personal health from sociobiological considerations.

What we may say at this point is that Boorse is quite right in emphasizing a 'species-typical' contribution but quite mistaken in restricting it to survival and reproduction. It is, without a doubt, this very restriction that encourages him to insist: "First, diseases are interferences with natural functions. Second, since the functional organization typical of a species is a biological fact, the concept of disease is value-free" ([3], p. 63). In terms of what has already been supplied, the following objections seem entirely fair. (a) Boorse cannot conclude from the fact that the functional organization typical of a species is thus-and-such that the concept of disease is value-free: the conceptual connection between disease and the statistical facts does not seem that clear. (b) whether Boorse's view of the relevant functional organization is or is not value-free depends decisively on what he means to specify as 'species-typical': for example, shifting patterns of aging that depend on *technological* and *social* innovations may affect our conception of what, relevantly, is 'species-typical'. It is entirely possible, for instance, that the strategic selection of temporal slices of the population of the species would affect substantially the species-typical regularities said to fix the conditions of disease and illness. Also, it seems quite impossible, given the radically different quality of life of different sub-populations of the race at different times and in different places, to make any sense of the 'species-typical contribution' of the parts and processes of the human body to survival and reproduction – without some (however innocent) *non*-statistically selected paradigm modes of functioning. But if ideological expectations affect our account of what is species-typical, then, clearly, functional organization cannot be restricted to merely biological considerations; hence, the concept cannot be value-free. (c) If functional organization relevant to health and disease were, as seems reasonable, expanded beyond considerations of survival and reproduction (whether addressed to the species or the individual – as Boorse prefers), it would be impossible to free our account of what is relevantly species-typical from ideological intrusion: alternative conceptions of health and disease would have to favor, variably, the differences in social expectation and technological achievement that characterize different human populations. But if this is so, then we see the infelicity of Boorse's condition (ii), that a disease is an internal state of an organism which "is not simply in the nature of the species". With the development of human technology and with the consequent changes in human expectations even with regard to biological functioning, it is quite impossible to pretend that what is "in the nature of the species" can be detailed in any straightforward way – certainly not with respect to survival and reproduction.

These considerations lead to an even more decisive issue. It is extraordinary that, in speculating about the nature of health and disease, the centrality of the concept of a person is so often overlooked. Symptomatically, Boorse states that he intends to "argue that the functional idea of health in physical medicine applies as straightforwardly to the mind as to the body", his intention is to show that the usual 'mental-health ideals' are "all of them . . . inappropriate to a theory of health", though something may be done to recover mental health 'after all' ([3], p. 62). What Boorse overlooks is simply that *physical health* itself is a condition ascribable to *human persons* and not merely to *human bodies*. He recognizes, correctly, that the adoption of materialism does not in itself entail the reduction of mental-health theory to physical-health theory ([3], pp. 61–68). But of course there is absolutely no compelling evidence that a reductive materialism is tenable.[9] In particular, if we think of human persons as culturally emergent creatures capable at least of speech and self-reference and of whatever culturally significant endeavors such capacities may permit, then it is quite impossible to suppose that what is taken to be *biologically* species-typical of *Homo sapiens* is sufficient to fix what is, on any thesis at all, relevantly typical of human persons. There is no question that all theories of mental health are concerned with providing norms suitable to human persons. What the ontological relationship between the human organism and the human person is need not concern us at this point.[10] What is essential, however, is that *if* persons as such are not characterizable merely as members of a biological species, then Boorse's condition (ii) is not only indecisive for a theory of mental health, it is indecisive for a theory of physical health [24]. Furthermore, *if* persons are culturally emergent beings and *if* there is no prospect of defending a cognitivism with respect to the norms of cultural life, then it is impossible to construct a theory of health appropriate to human persons that is either value-free or, if value-laden, free of the ideological divergences that characterize different societies. But if the intrusion of doctrinal values is unavoidable, then, first of all, we must radically revise our conception of the boundaries of medicine; and, secondly, we must reconsider the relationship between 'social maladaptation' and 'malfunctioning machinery' with a view to tolerating a confused area between medical and moral (or other social) values.

What has still to be demonstrated is that the considerations advanced do not entail a radical relativism or skepticism regarding medical values – whether directed to physical or mental health. But perhaps we should first collect the principal alternatives already scanned. Szasz rejects the pretensions of psychiatry as medicine, on the grounds that 'patients' are socially maladjusted and that maladjustment cannot rightly be subsumed under the models of physical medicine. Some, notably the behavior therapists, tend to construe psychiatry as a medical discipline but, at the same time, view relevant disorders in terms of social deviance or socially labelled maladjustment.[11] Heinz Hartmann holds that health is indeed a value but objectively confirmed for psychiatry as for physical medicine; and that, as 'technical' values, health values may be cleanly sorted from the personal or social values that a physician contingently adopts. Boorse holds that the usual mental-health values favored in psychiatry and psychotherapy are indefensible insofar (but only insofar) as they are construed in terms of social maladaptation and deviance rather than in terms of physiological or mental function; that the concept of physical health is value-free *and* that it is entirely possible to construct a theory of mental health that is equally value-free. The counterthesis being developed here is that all of these alternatives are equally mistaken; in particular, (i) that neither the criteria of physical nor of mental health are discoverable or value-free; (ii) that the functions assignable in human medicine are assignable to human persons, not merely to biological organisms, hence, they cannot be formulated without some attention to characteristic cultural endeavors; (iii) that the norms of human health cannot, therefore, be confined to what is merely (biologically) species-typical; (iv) that even with respect to the biologically species-typical, we characteristically concede that the typical (on any fair view) may yet be medically undesirable – diseased or ill; and (v) that the norms of health cannot, therefore, fail to be ideologically informed and infected. Admittedly, these conclusions seem to favor a dismal view of the objectivity of medicine. But there is a perfectly straightforward way of redeeming the professional venture and of reconciling the partial insights of all of the commentators scanned.

Consider that empirical evidence overwhelmingly confirms that human beings tend to favor such values as survival, reduction of pain and suffering, avoidance of incapacitation with respect to a wide range of projects, security and comfort of person and body, gratification of desires, and the like. There can be no complete table of such concerns, but there need not be. They may be called values in the relatively neutral sense that they are actually favored in the aggregate by the race. And they may be called prudential values in the joint sense that, so favored, they are essential to the preponderant part of the culturally determined values that different societies favor – though, notably, not to all such values, or at least not in any simple way; and that they are normally assumed to be sufficient to establish a minimal grade of rationality on the part of agents engaged in any particular significant endeavor. In suicide and in self-sacrifice and war and the like, it should be noted, prudential values are often waived or subordinated to other values or concerns thought to take precedence – without, however, necessarily entailing irrationality [19]. This shows at once the sense in which the specification of prudential values as somehow 'typical' of human persons cannot, however plausible it may be, be altogether freed of some minimal importation of normative values: which reminds us once again of the ambiguity of Boorse's conception of the 'species-typical' features of medicine.

Prudential values, furthermore, are merely determinable. Certainly, it is not true that, on the statistical evidence, there is some actual determinate length of life, determinate condition of reduction of pain, incapacitation, or insecurity or the like that is definitely either species-typical or typical of human persons. Only some culturally shaped doctrine, ideology, tradition, institution, or practice can be expected to give determinate form to the prudential values mentioned. Now, the interesting thing is that all of the substantive (frankly, normative) values by which human beings and human societies guide themselves – political, legal, religious, moral, and aesthetic – may fairly be construed as determinate specifications of prudential values or as superior values serviced more favorably by one determinate specification of prudential values than by another [19]. Both medicine and the law are, in this sense, ideologically variable articulations of certain selected prudential values. The admission of prudential values serves to confirm the conservative foundation of every reasonable system of medicine and law; and the admission that such values are only determinable serves to confirm that any adequate system of medicine or law (that is, any system capable of giving determinate direction over a comprehensive range of decisions and judgments) cannot escape incorporating the culturally variable convictions of societies at different levels of technological achievement and different levels of social expectation. The implicit relativism of this account is, therefore, saved from implausibility, since prudential values set a minimal constraint on cultural divergence.[12]

To return to our principal concern, the values of physical medicine are functionally specified in terms of determinate interpretations of the particular prudential values they are designed to service. Medicine is rather more fundamental than law, in the sense that the prudential values favored in the first constitute the most elementary conditions on which *any* of a very wide range of possible endeavors would depend: this is the point and the conservative objective of medicine; and this also explains why commentators like Boorse are persuaded that the discovery of the medically relevant physiological functions of the body is a straightforward business. A few characteristic citations from Boorse will, by this time, be entirely self-explanatory:

> ... a theory of health should be a description of how we are constituted and not how we would like to be ([3], p. 70). However we may disvalue neurosis and seek to eradicate it, we cannot call it unhealthy until we know that the mind is not supposed to work that way. It is in no way obvious, and requires empirical support, that what clinicians see in their offices are usually cases of biological dysfunction ([3], p. 71). Functions in physiology are species-typical contributions of a part to survival and reproduction ([3], p. 76).

Two themes invite a sharper focus. *If*, on the foregoing argument, prudential values must be assigned to human persons, then, although one may reasonably speak of biologically critical functions – thinking of the concept of health – those functions are so specified *relative to the prudential concerns of human persons.* Physical medicine tends to be conservative merely because, through the entire experience of the race, there have been relatively small changes in what has been judged prudentially pertinent in the condition and capacity of the human body. There have been *some* changes, however, as the changing conception of normal aging amply testifies. And it is entirely conceivable that further radical changes in the social organization of man, in the specialization of human contributions within enormously increased populations (say, approaching 30 billion individuals), in unheard-of technical innovations that could render hitherto unchangeable and therefore tolerated conditions of decline no longer beyond adjustment might well entail relatively radical changes in the concept of even the most basic physiological functions of the human organism. To see this is, of course, to see the inevitable ideological element in medicine. Furthermore, if this be conceded, then it becomes quite impossible to speak of a medically relevant "*description* of how we are constituted" or of straightforwardly knowing how "the mind *is supposed to work.*"

The second issue concerns the ambiguity of speaking of functions. The last citation from Boorse neatly shows the way: "Functions in physiology are species-typical contributions *of a part to* survival and reproduction" ([3], p. 76). It is easy to see that a part may be said to function so as to contribute to the way in which an organism happens to function, and yet lack *a proper function.* Boorse apparently wishes to disallow the distinction between *the* function of a part and *how* a part *happens* to function. This is the reason he maintains that certain organs function, happen to function – given the statistical evidence – to contribute in a species-typical way to survival and reproduction. A radical change in the statistical facts would entail a change in the characterization of the function of one's parts – hence, of disease and illness. But this leaves no conceptual ground for arguing that the way in which we happen to function (statistically, to be sure, but in a way that, as with pollution, may well be the result of our own purposive endeavors) may be diseased, ill, certainly undesirable [37, 24]. It appears anomalous that Boorse's doctrine does not provide for the medically relevant appraisal of changes in the human condition due to the human manipulation of the environment (and even of genetic factors): it seems preposterous to suppose that, given a powerful medical technology, the only objective judgments of disease and illness must be restricted, say, to what might restore an individual's physiological parts to whatever, contingently, happens (in some sense) to be the species-typical mode of functioning of those parts at the time. For example, it seems not unreasonable that medicine should invoke the (now temporarily) lost level of functioning of those parts as of an earlier time or the (now temporarily) not yet achieved but conceivable level of functioning of those parts of some technologically accessible future. It is hard to see how considerations of this sort are conceptually irrelevant to judgments of disease and illness.

These considerations, then, draw us to the following general conclusions: (a) that the determinate functions of survival and reproduction for individual humans cannot fail to be doctrinally specified within a given culture; and (b) that other determinate functions of the parts or processes of human persons are, relative to man's prudential interests, equally legitimately specified – however ideologically variable they may be – but not on purely biological grounds. *If*, then, medical values are selected values for persons and *if* persons cannot be characterized solely or primarily in biological terms, then (1) the *functions* medically assigned to the parts and processes of persons cannot but be ideologically skewed; (2) those functions cannot be biologically discovered; (3) *the* physiological function of the parts of the human organism cannot, consequently, be discovered either; and (4) we are bound to distinguish between how any of the parts and processes of human perons happen to function and how they ought to function, on some relevant theory of disease and illness. The point is that functions need not be open to discovery. The conservative nature of the conditions of biological change and prudential interest insures the relative invariance, on dialectial rather than scientific grounds, of the values of physical medicine. In the same spirit – though, noticeably, not with an equally obvious neutrality – the functional norms of mental health are specified. It is true, of course, that the values of mental health are infinitely more susceptible to ideological manipulation than the values of physical health. But that cannot be more than a matter of degree. It cannot be a matter of difference in kind. Still, such differences of degree are sufficient for all the quarrels of the world.[13]

Temple University
Philadelphia, Pennsylvania

NOTES

[1] *Cf.* Rieff [29]; also Wollheim [34], especially 'Freud's Anthropomorphism' by Thomas Nagel and 'Freud's Neurological Theory of Mind' by Robert C. Solomon; *cf.* also Freud [6], including *Scientific Project.*

[2] For a more detailed discussion of the logical aspects of recommending a change in medical classification relative to Freud, see my article [13]. A fuller discussion of Szasz's position is given in [12].

[3] On challenges to Freud's conception of the normal sexual development of women, *cf.* Mitchell [27], Strouse [31], and Klaich [10].

[4] The issue is discussed at length in my book [15], Chapter 1.

[5] With respect to the literature of mental health, *cf.* Jahoda [9]; Wooten [36]; and my book [12]. With respect to physical health, *cf.* my article [24].
[6] *Cf.* Margolis [11]. The argument is directed against Stuart Hampshire [7], who has offered the most thoroughgoing Platonic interpretation of classification that we have.
[7] See Boorse ([2], p. 61). He gives precisely the same formula in ([3], p. 63).
[8] See Boorse [3], p. 72. One must consider the complications generated by the phenomenon of 'balanced' genetic load, which requires us to think of the health of a population in terms of survival and reproduction, but only on the condition of tolerating, within the normal span of health, sub-populations (and their members) that may be distinctly weaker with respect to survival but which may supply a genetic resource necessary for the survival of the entire population. I have discussed this more pointedly in [24]. *Cf.* also Dobzhansky [4].
[9] I have pursued the difficulties of reductive materialism in a number of different ways. *Cf.*, for example, Margolis [14, 16, 17, 21, 22].
[10] My own thesis, resisting reductionism and dualism, is that persons are *embodied* in the biological members of Homo sapiens and emerge culturally with the training of such creatures in speech and other cultural skills. I hold that all cultural entities – persons, works of art, language, machines – are similarly embodied entities; the details need not concern us at this point, since the distinction of the human person is all that is required. But for a fuller account of the conceptual issues developed for works of art, not persons, *cf.* Margolis [20], and Margolis [25]. The application to persons appears in my book [26].
[11] Boorse cites the position of Redlich [28]: 'behavior disorder' refers to the presence of certain behavioral patterns – variously described as abnormal, subnormal, undesirable, inadequate, inappropriate, maladaptive, or maladjusted – that are not compatible with the norms and expectation of the patient's social and cultural system; also, the position of Ullmann [33]: "Maladaptive behavior is behavior that is considered inappropriate by those key people in a person's life who control reinforcers."
[12] Again, the thesis is developed for all the principal areas of social concern in my book [19]. The logical properties of this sort of relativism – what I term 'robust relativism' – are developed in my article [23]. The concept is applied, there, to the interpretation and the evaluation of works of art, but the properties specified are entirely general.
[13] I have benefitted considerably from several exchanges by letter with Boorse. I believe I have avoided some errors regarding his views thereby, but I have also apparently confirmed my disagreement with him.

BIBLIOGRAPHY

1. Baker, R. and Elliston, F. (eds.): 1975, 'The Question of Homosexuality', in *Philosophy and Sex*, Prometheus Books, New York, pp. 288–305.
2. Boorse, C.: 1975, 'On the Distinction between Disease and Illness', *Philosophy and Public Affairs* 5, 49–68.
3. Boorse, C.: 1976, 'What a Theory of Mental Health Should Be', *Journal for the Theory of Social Behaviour* 6, 61–84.
4. Dobzhansky, T.: 1962, *Mankind Evolving*, Yale University Press, New Haven.
5. Feinstein, A.: 1967, *Clinical Judgment*, Williams and Wilkins, Baltimore.

6. Freud, S.: 1954, in Marie Bonaparte, Anna Freud, and Ernst Kris (eds.), *The Origins of Psychoanalysis*, translated by Eris Mosbacher and James Strachey, Basic Books, New York.
7. Hampshire, S.: 1959, *Thought and Action*, Chatto and Windus, London.
8. Hartmann, H.: 1960, *Psychoanalysis and Moral Values*, International Universities Press, New York.
9. Jahoda, M.: 1958, *Current Concepts of Positive Mental Health*, Basic Books, New York.
10. Klaich, D.: 1974, *Woman Plus Woman*, Simon and Schuster, New York.
11. Margolis, J.: 1963, 'Classification and the Concept of Goodness', *Australasian Journal of Philosophy* **41**, 182–185.
12. Margolis, J.: 1966, *Psychotherapy and Morality*, Random House, New York.
13. Margolis, J.: 1969, 'Illness and Medical Values', *The Philosophy Forum* **8**, 55–76.
14. Margolis, J.: 1971, 'Difficulties for Mind-Body Identity Theories', in Milton Munitz (ed.), *Identity and Individuation*, New York University Press, New York, pp. 213–233.
15. Margolis, J.: 1971, *Values and Conduct*, Clarendon, Oxford.
16. Margolis, J.: 1973, 'The Perils of Physicalism', *Mind* **82**, 566–578.
17. Margolis, J.: 1974, 'Reductionism and the Ontological Aspects of Consciousness', *Journal for the Theory of Social Behavior* **4**, 3–16.
18. Margolis, J.: 1975, 'Moral Cognitivism', *Ethics* **85**, 136–141.
19. Margolis, J.: 1975, *Negativities: The Limits of Life*, Charles Merrill, Columbus.
20. Margolis, J.: 1980, *Art and Philosophy*, Humanities Press and Harvester Press, New York and Hassocks.
21. Margolis, J.: 1976, 'On the Ontology of Persons', *New Scholasticism* **50**, 73–84.
22. Margolis, J.: 1976, 'Countering Physicalistic Reduction', *Journal for the Theory of Social Behavior* **6**, 5–19.
23. Margolis, J.: 1976, 'Robust Relativism', *Journal of Aesthetics and Art Criticism* **35**, 37–46.
24. Margolis, J.: 1976, 'The Concept of Disease', *The Journal of Medicine and Philosophy* **1**, 238–255.
25. Margolis, J.: 1977, 'The Ontological Peculiarity of Works of Art', *The Journal of Aesthetics and Art Criticism* **36**, 45–50.
26. Margolis, J.: 1978, *Persons and Minds: The Prospects of Nonreductive Materialism*, D. Reidel Publ. Co., Dordrecht, Holland.
27. Mitchell, J.: 1974, *Psychoanalysis and Feminism*, Pantheon, New York.
28. Redlich, F. and Freedman, D. X.: 1966, *The Theory and Practice of Psychiatry*, Basic Books, New York.
29. Rieff, P.: 1961, *Freud: The Mind of the Moralist*, revised, Anchor Books, Garden City, N.Y.
30. Sarbin, T.R.: 1969, 'The Scientific Status of the Mental Illness Metaphor', in Stanley C. Plog and Robert B. Edgerton (eds.), *Changing Perspectives in Mental Illness*, Holt, Rinehart, and Winston, New York, pp. 9–31.
31. Strouse, J. (ed.): 1974, *Women and Analysis*, Grossman Publishers, New York.
32. Szasz, T. S.: 1961, *The Myth of Mental Illness*, Hoeber-Harper, New York.
33. Ullmann, L. P. and Krasner, L.: 1965, *Research in Behavior Modification: New Developments and Implications*, Holt, Rinehart, and Winston, New York.

34. Wollheim, R. (ed.): 1974, *Freud: A Collection of Critical Essays*, Anchor Books, Garden City, N.Y.
35. Wolpe, J.: 1973, *The Practice of Behavior Therapy*, 2nd ed., Pergamon Press, New York.
36. Wootten, B.: 1959, *Social Science and Social Pathology*, Macmillan, New York.
37. Wright, L.: 1973, 'Functions', *Philosophical Review* **82**, 130–168.

MICHAEL S. MOORE

LEGAL CONCEPTIONS OF MENTAL ILLNESS

I

By attaching legal consequences to conduct, the law necessarily regards individuals as responsible agents. For example, in holding individuals civilly and/or criminally liable for actions harmful to others, the law necessarily presupposes that it applies to individuals who are responsible, in the sense of accountable, for their actions; in enforcing certain voluntary acts of individuals, as in the law of property, contracts, wills, or marriage, the law must presuppose that the agents whose acts they are, are responsible in the sense of being capable of looking out for their own best interests. Mental illness has for centuries been a concept dealt with by the law because it negates, in some way or another and to some degree, the basic postulate of responsibility on which the law rests.

Broadly speaking, mental illness figures into two kinds of legal contexts. These are perhaps best distinguished by first distinguishing two senses of the word 'responsibility': a retrospective sense, and an overall sense ([4, 19], pp. 17–45; [25], p. 211). Retrospectively, we hold people responsible *for* certain events in the past. We make such ascriptions of responsibility based on a host of criteria, involving concepts of causation, intention, voluntariness, and action, and matters of justification or excuse. In the overall sense, we speak of persons being responsible in the sense that they are the sort of people who can be counted on to make 'responsible' decisions. If an agent is mentally ill, his responsibility in both senses is in question. Taking the retrospective sense first, in the criminal law, he may be adjudged legally insane and thus not the subject of punishment. In the law of torts, at least if his illness is of a sudden sort, an agent may be relieved of responsibility for harm he has caused in that he may not have to pay damages in a civil suit by the injured party; in constitutional law, it may constitute cruel and unusual punishment in violation of the federal constitution to punish an individual for acts brought on by his mental illness. In each case, he may be relieved of legal responsibility for some harm that he has caused.

With regard to the overall sense of responsibility, a mentally ill person may not be regarded as sufficiently responsible to be able to manage his own

B. A. Brody and H. Tristram Engelhardt, Jr. (eds.), Mental Illness: Law and Public Policy. 25–69.

property, to make his own contracts, wills or marriage, to be granted custody of his children in a divorce setting, or even to make fundamental decisions such as those concerning his liberty or whether he wishes to be medically treated in one way as opposed to another. In each of these cases, the law does not regard the mentally ill person as sufficiently responsible to have his decisions accorded the same legal consequences as those accorded similar decisions made by others.

The first legal context in which the concept of mental illness is utilized, is one in which responsibility *for* some harm is in question (i.e., torts and criminal law). Here, mental illness is relevant as a moral and legal excuse. The second context in which the concept of mental illness is utilized is one in which one's overall responsibility is in question, so that the state is willing to make a paternalistic decision regarding the best interest of the individual in question.

These two legal contexts do not exhaust the ways in which mental illness may come before the courts. When the state civilly commits mentally ill persons on grounds of dangerousness to others (as opposed to the paternalistic grounds of dangerousness to self or the need for treatment, the other two traditional grounds of commitment), mental illness is relevant in a way different from the two responsibility assessments just discussed. Because of this difference, however, it has been persuasively argued that civil commitment on grounds of dangerousness to others ought not involve the concept of mental illness at all, and that any individual, mentally ill or not, about whom a reliable prediction of dangerousness can be made, should be detained. The concept of mental illness enters the law at other miscellaneous points as well, including problems arising: (1) in workmen's compensation law where the issue is whether mental illnesses qualify as compensable injuries; (2) in an insurance context where insurance clauses regarding illness must be construed as to their inclusion of mental illness; and (3) in malpractice suits in tort law against doctors for failure to diagnose certain mental conditions as illnesses. The main point is preserved, however; legal definitions of mental illness have by and large arisen from one of the two contexts above described, contexts in which the responsibility of an agent is in question. The concept of mental illness that the law has developed must be seen in this light.

I will not survey each of these branches of the law for differences in the kind of conceptions of mental illness that courts and legal scholars may have utilized. Instead, I shall focus on one such area of law, that which is the central focus of the context I called assessment of retrospective responsibility: the defense of insanity in the criminal law. Legal insanity in some form has been an excuse from criminal responsibility for centuries. Its proper definition has been extensively debated, by psychiatrists as well as lawyers; it is by far the richest in history of any legal attempt to deal with the concept of mental illness. It is thus an appropriate context in which to assess the contributions to the contemporary debate about the status of the concept of mental illness that law and lawyers might make.

II. BRIEF SKETCH OF THE INSANITY TESTS

A. Brief History of the Definition of Legal Insanity

1. Pre-Nineteenth-Century Definitions

Insanity in some form was an excuse in most ancient systems of law. Ancient Mohammedan law applied punishment only to "individuals who have attained their majority", and "who are in full possession of their faculties" ([8], p. 311). This analogy of the insane to children exists as well in ancient Hebraic law, which recognized that deaf-mutes, idiots and minors were not responsible for their actions ([17], pp. 501–502). Roman law, at least during the latter period of the Empire, also relieved the mentally ill of responsibility for wrongful actions, again drawing the analogy of insane persons explicitly to children [6].

Two analogies dominate the earliest ideas about the insanity defense in Anglo-American law. The first of these is again the analogy to children. Sir Matthew Hale, Chief Justice of the Court of the King's Bench, in his *Select Pleas of the Crown* expressed the view that the best measure of legal insanity was whether or not the accused "hath yet ordinarily as great understanding as ordinarily a child of 14 years hath" [50]. This explicit analogy to the responsibility of children was implicit in a much earlier development in Anglo-American law, the development of the good and evil test, which was later to influence the M'Naghten test of insanity in the nineteenth century. The good and evil test is thought to have been taken out of Genesis, where knowledge of good and evil likens a person to God: in the garden the serpent says to the woman, "You will be like God, knowing good and evil". And God replies "behold, the man has become like one of us knowing good and evil" (Genesis 2:16–3:1, 3:4–5, 3:22–23). This language from the Bible was picked up at least as early as the fourteenth century (as the test of when a child became a responsible person). In 1313 it was held that "an infant under the age of seven years, though he be convicted of felony shall go free of judgment, because he knoweth not of good and evil" [67]. It was this explicit analogy to the conditions under which children could be regarded as responsible agents, that was behind the infusion of this language into a definition of legal insanity [40].

The second famous analogy of early Anglo-American law was to animals. As early as the thirteenth century, Bracton, in the *De Legibus Et Consuetudinibus Angliae,* the earliest comprehensive treatise on English law, thought that madmen were "not greatly removed from beasts for they lack reasoning" ([8], p. 82). Lord Coke, in Beverly's case in 1603, quoted what he took to be Bracton's analogy with approval (*Beverly's Case,* 4 Coke 1236, 1246 (1603)). It was in Arnold's case, however, in the early eighteenth century that the analogy to animals became well-entrenched in English law. Justice Tracy, presiding at the trial of Edward Arnold, said that in order to avail himself of the defense of insanity, "a man must be totally deprived of his understanding and memory, so as not to know what he is doing, no more than an infant, a brute, or a wild beast" (Rex v. Arnold, 17 How. St. Tr. 695, 764 (1724)).

2. *The M'Naghten Test*

In 1843 Daniel M'Naghten shot and killed Edward Drummond, private secretary to the Prime Minister, Robert Peel. M'Naghten was under the delusion that he was being persecuted by a host of individuals throughout England and Scotland, including, he thought, the Prime Minister. At this trial he successfully raised the defense of insanity. Given the political unrest at the time, the suspicion that M'Naghten was merely feigning his illness, and the fact that Queen Victoria herself had recently had an assassination attempt made against her life for which the assailant was excused by reason of insanity, the Queen was outraged at the result of excusing M'Naghten from criminal responsibility. The House of Lords accordingly asked the judges to appear before them as a group and explain the proper tests of criminal insanity. The M'Naghten rules originated in the judges' answers to questions put to them by the House of Lords. The much quoted and operative language of the judges' answers was that:

> To establish a defense on the grounds of insanity, it must be conclusively proved that, at the time of the committing of the act, the party accused was laboring under such a defect of reason, from the disease of the mind, as not to know the nature and quality of the act he was doing; or if he did know it, that he did not know what he was doing was wrong. (*Regina v. M'Naghten,* 10 Clark and F. 200, 8 Eng. Rep. 718 (1843))

The M'Naghten test quickly became the leading test for insanity in

England and America. It is still the exclusive test for insanity in a majority of American states, and part of the test of insanity in almost all other state jurisdictions ([23], p. 45). The essential elements of the test, about which so much has been written, are three: first, that one suffer from a defect of reason stemming from a disease of the mind; secondly, that one lack the knowledge of what one is doing; and thirdly, that one lack moral knowledge, the knowledge of good and evil, which originated in the responsibility test for children.

3. The Irresistible Impulse Test

One of the persistent criticisms of the M'Naghten test in the nineteenth century, and continuing to this day, is that it does not excuse from criminal responsibility a large enough class of persons. More specifically, the M'Naghten test was thought not to excuse those mentally ill persons who knew what they were doing and that it was wrong, but who nonetheless because of their mental illness did not have the ability to control their behavior. The distinction often made was between cognitive incapacity and volitional incapacity. The irresistible impulse test was formulated as a response to this criticism of M'Naghten. Although there is no one case giving a classic definition of legal insanity as irresistible impulse, one of the leading cases defined the defense as follows:

> Did he know right from wrong, as applied to the particular act in question . . . if he did have such knowledge, he may nevertheless not be legally responsible, if the two following conditions concur: (1) if, by reason of the duress or such mental disease, he had so far lost the power to choose between the right and wrong, and to avoid doing the act in question, as that his free agency was at the time destroyed; (2) and if, at the same time, the alleged crime was so connected with such mental disease, in the relation of cause and effect, as to have been the product of it solely. (Parsons v. State, 81 Ala. 577, 597, 2 So 854, 866–67 (1887).)

As Abraham Goldstein has pointed out ([23], pp. 67–79), the common label for this test – 'irresistible impulse' – is misleading. The essential notion of the test is that of a person who, because of his mental illness, has lost the power to control himself. The test might more properly be called a 'loss of control' test.

4. The American Law Institute's Definition of Legal Insanity

In the 1950's the scholars and judges comprising the American Law Institute proposed a Model Penal Code for adoption by state and federal jurisdictions.

Section 4.01 of that Code included a new definition of legal insanity, which was as follows:

(1) a person is not responsible for criminal conduct if at the time of such conduct as the result of mental disease or defect he lacks substantial capacity either to appreciate the criminality of his conduct or to conform his conduct to the requirements of law. (2) the terms 'mental disease or defect' do not include an abnormality manifested only by repeated criminal or otherwise antisocial conduct [33].

This test, or some variant of it, has been adopted within the last fifteen years by all of the federal courts of appeals (*United States v. Brawner,* 471 F2d 969 (D.C. Cir., 1972)).

5. The New Hampshire and Durham Experiments

As I shall describe in much more detail shortly, New Hampshire and the District of Columbia (between 1954 and 1972) adopted as their criterion of legal insanity the following: "an accused is not criminally responsible if his unlawful act is a product of mental disease or of mental defect" (*Durham v. United States,* 214 F2d 862 (D.C. Cir., 1954)). This test is traditionally analyzed as having two elements: first, that the accused be suffering from a mental disease or mental defect ('mental disease' being used in the rule as a synonym of 'mental illness' [*Carter v. United States*, 252 F2d 608, 617 (D.C. Cir., 1957)]); and secondly, that his criminal act be the product of that diseased or defective condition.

B. Classification of Insanity Definitions: The Moral and Psychiatric Paradigms Underlying the Insanity Tests

Writers in the area of legal insanity have often been confused between two quite distinct concepts: legal insanity and mental illness. Although I shall argue that the two ought to be equated and legal insanity defined as mental illness (at least as the mental illness of a popular moral paradigm which I shall analyze in part V of this paper), it is at this stage nonetheless helpful to keep the two distinct [66].

Some order can be brought into these tests by observing the relationship between mental illness and legal insanity. To be legally insane is to be excused from criminal responsibility. Each of these definitions of legal insanity thus is a test determining when an accused is or is not responsible in the criminal law. Each of the tests, it will be observed, has mental illness or some related concept as one of the elements. With the exception of *Durham* and the New Hampshire rule, however, the definitions of legal insanity typically do not equate insanity and mental illness, but have mental illness as only one element, one criterion, of determining when someone is legally insane. To be legally insane under M'Naghten, irresistible impulse, or the American Law Institute test requires, in addition, that certain other criteria be met. Each of these tests is best viewed as an attempt to adjust the then prevailing view about mental illness to well-established moral and legal paradigms of excuses from responsibility. To see the moral paradigms involved requires a brief detour.

In criminal law as in morals, two general sorts of conditions excuse: ignorance that is not itself culpable, and compulsion. Such excuses are distinct from other modes of defeating the ascription of legal or moral responsibility known as justification, such as in self-defense. These two moral excuses are as old as Aristotle[1] and are embodied in contemporary criminal law. If one makes a mistake of fact about a material element of a crime, (e.g., one believes some substance to be harmless food coloring when, in fact, it is poison), one is not held liable to punishment; similarly, those who act under duress (threats by others), or those who act under the 'internal compulsion' of severe emotional states arising out of the victim's provoking acts, are also excused or partially excused from punishment.

It is these two traditionally excusing conditions that have been adopted by lawyers and judges and incorporated into the traditional definitions of legal insanity above set forth. There are thus basically two kinds of traditional insanity tests: those based on the ignorance of the mentally ill accused person; and those based on some notion of his being compelled to act as he did.[2]

The M'Naghten formulation quite obviously is of the first type, which turns on the ignorance of the accused about what he is doing or its moral status. The M'Naghten opinion was written, and its rules adopted by other courts, at a time when delusions were thought to be the prominent symptoms of mental illness. Indeed, many judges and lawyers thought that the presence of delusions was the only criterion for being mentally ill.[3] It is thus not surprising that a test should have evolved which combined this conception of mental illness – as delusion – with the long-existing moral paradigm that ignorance was an excuse. While the old language about moral knowledge was retained, the rationale for the language was forgotten; as transformed by the M'Naghten rules, the knowledge required was not the general knowledge that was the measure of when a child or an insane person has the mental maturity to be treated as a responsible agent, but was rather the knowledge relevant to determining whether one could avail oneself of the traditional excuse of ignorance of fact or law, namely, factual and moral knowledge about the particular act.

The irresistible impulse test also represents this fitting of psychiatric insight into an already existing paradigm of moral excuse. It was principally Isaac Ray's criticism of the M'Naghten test in the nineteenth century, to the effect that delusion was not the only symptom of mental illness, that persuaded courts to frame an alternate definition of legal insanity around the other existing paradigm of moral excuse, that of internal compulsion.[4] Whether one was ill in the sense of having lost his ability to control himself, a 'volitional incapacity' as opposed to a 'cognitive impairment', was the essential question under irresistible impulse. Such a test received further impetus with the prevalence of psychoanalytic theories in the 1920's, whose talk of instinctual drives, energies, and forces all at least seemed to add to the idea that mentally ill persons are in some fundamental sense compelled to act as they do and are thus not responsible ([23], p. 67).

The American Law Institute test is simply a rewording of each of these aspects of the nineteenth century tests of legal insanity, and their conjunction into one single test. Instead of focusing on 'knowledge' as in M'Naghten, the ALI test talks of 'appreciating' the criminality of conduct; in place of irresistible impulses and inabilities to control, the ALI talks of a lack of substantial capacity to conform conduct to requirements of law. The moral paradigms invoked are wholly the same.

Distinct from all of these tests is the New Hampshire and *Durham* formula of legal insanity, for this definition equates legal insanity with mental illness. Rather than incorporating long-existing paradigms of moral excuse from other areas of criminal law into a definition of insanity, the New Hampshire and *Durham* definitions regarded mental illness as itself an excusing condition, even if not accompanied by ignorance or compulsion. Justice Doe, the originator of the New Hampshire test in the nineteenth century, thought that he was returning to the ancient ideas about legal insanity; namely, that mental illness itself excuses. In this respect he was correct; the status of being mentally ill, just like the status of being a child, itself excused one from responsibility. Unfortunately, however, Justice Doe of New Hampshire and Judge Bazelon of the District of Columbia Court of Appeals were both also heavily influenced by the psychiatric theories of mental illness with which they were contemporary. Their interpretations of their respective definitions of legal insanity were thus more influenced by the prevailing psychiatric theory than by the ancient paradigm to which on occasion Justice Doe thought he was returning.

The influence of psychiatry and contemporary opinions about mental illness on the judges who wrote the New Hampshire and *Durham* tests is complicated and will be treated separately below. The history of the origins and administration of the insanity test in those jurisdictions not only illuminates the nature of psychiatric influence on the law's definitions, but also, because of the equation of legal insanity with mental illness, reveals that such tests involved the law directly in search of an adequate definition of mental illness.

III. THE NEW HAMPSHIRE AND *DURHAM* EXPERIMENTS: THE LEGAL/PSYCHIATRIC SEARCH FOR A DEFINITION OF MENTAL ILLNESS.

A. New Hampshire

In 1838 a little-known physician in Maine, Isaac Ray, published *A Treatise on the Medical Jurisprudence of Insanity* [45]. The book quickly became the authority in its field and remained the seminal work in forensic psychiatry in the nineteenth century. The book was relied on heavily, for example, by the trial judge in M'Naghten's case (although ignored in the formulation of the M'Naghten rules in the House of Lords) ([8], pp. 100–102), [15]. Ray's work was also highly influential on those early courts in America that adopted an irresistible impulse test ([45], Section 187). More pertinent to our immediate concern, however, is Ray's direct and well-documented influence on the development of the definition of legal insanity in New Hampshire.

Ray, like most nineteenth-century psychiatrists, was convinced that mental illness was in essence a brain disease, and that lesions in the brain would eventually be discovered as the cause of mental disorders:

> It is undoubted truth that the manifestations of the intellect and those of the sentiments, propensities, and passions, or generally of the intellectual and affective powers are connected with and dependent upon the brain. It follows, then, as a corollary, that abnormal conditions of these powers are equally connected with abnormal conditions of the brain; but this is not merely a matter of inference. The dissections of many eminent observers ... have placed it beyond a doubt; and no pathological fact is better established – though its correctness was for a long while doubted – than that deviations from the healthy structure are generally presented in the brains of insane subjects. ([45], Section 31)

This view of mental illness was eventually accepted by Justice Doe of the New Hampshire Supreme Court.

Justice Doe early in his career sought out the views of Ray on insanity, via the good offices of a Dr. Tyler of the Harvard Medical School. The three entered into a lengthy correspondence on the proper definition of legal insanity. Early in the correspondence, Doe asked Tyler:

> Is it now the settled opinion of the scientific world that insanity is only a physical disease, or the result of physical disease? ([47], p. 187)

If the answer were yes, Doe went on,

> Why should the court ever say to a jury more than this in cases of alleged mental aberration or active insanity: 'If the disposition of property was the offspring of or was caused or affected by mental disease, then it is not the will of the testator – the result of disease cannot have any effect in law.' ([47], p. 187)

Both Doe and Ray believed mental illness to be a physical disease. From this premise, it was self-evident to Justice Doe that the following proper definition of legal insanity should be eventually adopted in New Hampshire:

> If the homicide was the offspring or product of mental disease in the defendant he was not guilty by reason of insanity.[5]

The reasoning to this conclusion was based on a confusion continuing to this day: that if physical causes of behavior are discovered, the actor is *ipso facto* not responsible for that behavior. Doe thought that if the abnormal physical condition of a defendant's brain caused him to commit the criminal act, then necessarily his will, his power to choose, was extinguished – i.e., it was not his act:

> For if the alleged act of a defendant was the act of his mental disease it was not in law his act, and he is no more responsible for it than he would be if it had been the act of his involuntary intoxication or of another person using the defendant's hand against his utmost resistance . . . when a disease is the propelling, uncontrollable power, the man is as innocent as the weapon – the mental and moral elements are as guiltless as the material. If his mental, moral and bodily strength is subjugated and pressed to an involuntary service, it is immaterial whether it is done by disease or by another man or a brute or any physical force of art or nature set in operation without any fault on his part. (*State v. Pike*, 49 N.H. 399, 442 (1869))

The confusion inherent in this chain of reasoning stems from the nature of the language we use to make responsibility ascriptions, in morals as well as law. As is well known, the words with which we describe human actions imply the causal agency of the actor; to be an action at all (as opposed to a mere bodily movement), we must be able to assert that the agent (or the self, the mind, the conscious will, or what have you) was causally responsible for the bodily movements. If we are told that something else caused the bodily movement, then it *seems* that the agent was not causally responsible – for if some other set of conditions, such as an abnormal condition of the brain, was sufficient, that seems to imply that nothing else, such as the will of the actor, was even necessary.[6]

The problem with such a view lies in the failure to see the possibility of there being differing sets of equally sufficient conditions existing to cause the same event. To say that a bodily movement is the product of an abnormal condition of the brain does not preclude one from describing that movement as an action performed by an agent for reasons. We have two vocabularies – that of movement and mechanical causation, and that of actions and reasons. Merely because scientists may discover lesions in the brain is not to preclude the application of the language of action and reasons. Mechanically caused movements may nonetheless be intelligent actions [14].

It is the language of actions and reasons with which the law deals. In ascribing responsibility, the law adopts with little change the conditions we all use to ascribe moral responsibility in everyday life, conditions framed in the language of action and intentionality. Once one perceives the logical independence of this language from the language of natural science – the language describing brain pathology causing bodily movements – then assumptions such as those of Doe and Ray become irrelevant to the proper definition of legal insanity. Mental illness may be a physical disease – that debate is certainly still alive – but for legal and moral purposes, the outcome is irrelevant. If mental illness excuses, it is not because it is the name of an as yet unknown physical cause.

Despite their today-questioned assumptions that mental illness was brain disease, and despite their further erroneous assumption that because of this mental illness necessarily excuses, Doe and Ray were on the right track. For what they perceived to be inadequate about the M'Naghten test was its failure to recognize the excusing nature of mental illness itself. Although for the wrong reason (physical causation), Doe and Ray believed that legal insanity should be equated with mental illness because mental illness itself excuses. In that latter belief they were correct.

Their assumption that mental illness was a brain disease led them, however, into a further error for which they have been amply criticized by a century of legal scholars. Because of the assumed physical nature of mental illness, they thought that mental illness was a scientific concept and that accordingly, those scientists on the frontiers of investigating it – namely, psychiatrists – were the ones to inform the court on whether or not it was likely that a particular defendant did or did not have this abnormal brain condition. Again and again in their correspondence, in Doe's legal opinions and in Ray's writings, they reiterated the notion that the presence of mental disease was a fact to be testified to by medical experts:

Insanity . . . is the result of a certain pathological condition of the brain . . . and the tests and symptoms of this disease are no more matters of law than the tests or symptoms of any other disease in animal or vegetable life. (*Boardman v. Woodman*, 47 N.H. 120, 148 (1865) (Doe in dissent))

The law does not define disease – disease is so simple an expression that the law need go no further. What is a diseased condition of mind is to be settled by science and not by law – disease is wholly within the realm of natural law or the law of nature ([47], p. 187).

And finally:

The legal profession, in profound ignorance of mental disease, have assailed the superintendents of asylums, who knew all that was known on the subject, and to whom the world owes an incalculable debt, as visionary theorists and sentimental philosophers attempting to overturn settled principles of law; whereas in fact the legal profession were invading the province of medicine, and attempting to install old exploded medical theories in the place of facts established in the progress of scientific knowledge. (*State v. Pike*, 49 N.H. 399, 442 (1869))

The result of this view of the matter was clear: psychiatrists would have the authoritative voice on who was legally insane and thus to be excused from criminal responsibility.[7] The problem with such a result has been restated many times. The criteria of legal responsibility are for the law to settle. Even if mental illness is equated with legal insanity, the definition of it necessarily is a legal matter. It would be a pure coincidence if the concept of mental illness adopted by psychiatrists for therapeutic purposes were the same as the concept suitable to isolate the class of offenders who ought, consistently with the purposes of punishment, to be excused. Indeed, the point is quite general: the law must define legal concepts for itself in light of legal purposes. The law cannot simply adopt a concept developed by psychiatrists for therapeutic purposes, nor for that matter any concept developed by any social scientists for explanatory purposes. The purposes of the law in question must govern the definition of any term appearing in that law; no other discipline's conceptualization can safely be adopted and plugged into a legal formula.

The psychiatrists at the State Hospital in New Hampshire had the good sense to perceive that they could not have been asked under the New Hampshire test to plug their therapeutic notion of mental illness into the formula defining legal insanity. In fact, the concept of mental illness they used in their testimony about criminal defendents was not the same as the concept they used in classifying their patients for treatment purposes. They had to take upon themselves the task Judge Doe and his successors on the New Hampshire bench should have undertaken: namely, to give a separate, legal definition of mental illness as a legally excusing condition ([46], p. 19).

B. The Durham Experiment

This fundamental lesson had to be relearned when the Court of Appeals for the District of Columbia decided *Durham v. United States* in 1954 (*Durham v. United States*, 214 F2d 862 (D.C. Cir., 1954)). The *Durham* rule regarding insanity, thought by the court to be "not unlike that followed by the New Hampshire court" (*ibid*, at 874) was also that "an accused is not criminally responsible if his unlawful act was the product of mental disease or mental defect" (*ibid*, at 874–875). Unwittingly the *Durham* court followed in the steps of the New Hampshire court, both in relying initially on a medical conception of mental illness and in accepting the seeming consequence of that conception; namely, that causation was the real issue involved in the insanity defense. Each aspect of the court's reliance on the psychiatric paradigm of mental illness is pursued separately below.

1. Defining Mental Disease

Durham was decided explicitly to facilitate psychiatrists in placing their knowledge before the court, which they felt they could not do under the M'Naghten test. The influential Group for the Advancement of Psychiatry had earlier written a preliminary version of its report on criminal insanity, cited and relied upon in the *Durham* opinion. This report complained about "a barrier of communication which leaves the psychiatrist talking about 'mental illness' and the lawyer talking about 'right and wrong' " [24]. The test proposed by the committee, and in essence adopted in *Durham*, allowed psychiatrists to testify directly to the presence or absence of mental disease because the test was framed in terms of mental disease itself.

The problem, however, was the same as that which arose in New Hampshire almost a century earlier: not every medically recognized mental disease could have been intended. Something more restrictive must have been intended by the phrase "mental disease or defect" in the *Durham* rule.

This problem became particularly glaring with regard to sociopaths, a diagnosis which had been applied to Monte Durham himself. Shortly after the *Durham* decision, the staff at St. Elizabeth's Hospital, which was composed of those psychiatrists most often called to testify in the District of Columbia criminal cases, made a policy decision that sociopathic or psychopathic personality disturbances would not be regarded as mental illnesses within the meaning of the *Durham* rule. Psychiatrists from St. Elizabeth's Hospital thereafter so testified in District of Columbia cases. Three years later, however, a weekend staff meeting changed the St. Elizabeth's policy, and decided that henceforth, psychopathic or sociopathic personality disturbance would be considered a mental disease for legal purposes. The Court of Appeals for the District of Columbia deferred to this psychiatric judgment, granting a new trial in one case involving a sociopathic individual because, having been tried before the weekend change in classification by the psychiatrists, he was deprived of "new medical evidence ... on an issue vital to his defense", namely, whether he was mentally ill (*Blocker v. United States*, 274 F2d 572, 573 (D.C. Cir., 1959)). As Warren Burger, then a Circuit Judge who participated in that decision, later noted:

> We tacitly conceded to St. Elizabeth's Hospital the power to alter drastically the scope of a rule of law by a weekend change of nomenclature ... (*Blocker v. United States*, 288 F2d 853, 860 (D.C. Cir., 1961)).

This illegitimate transfer to psychiatrists of the power to decide the meaning of a legal rule on criminal responsibility resulted directly from the assumption of the District of Columbia judges that 'mental illness,' as used in the rule, was the same concept as that used in medicine. This assumption, implicit in *Durham* itself, had been made quite clear shortly after *Durham* was decided, when the Court of Appeals held that:

> Mental 'disease' means mental illness. Mental illnesses are of many sorts and have many characteristics. They, like physical illnesses, are the subject matter of medical science ... Many psychiatrists had come to understand there was a 'legal insanity' different from any clinical mental illness. That of course was not true in a juridical sense. The law has no separate concept of a legally acceptable ailment which *per se* excuses the sufferer from criminal liability. The problems of the law in these cases are whether a person who has committed a specific criminal act ... was suffering from a mental disease, that is, from a medically recognized illness of the mind ... (Carter v. United States, 252 F2d 608, 617 (D.C. Cir., 1957)).

Perceiving that surely not every medically recognized illness of the mind excuses from criminal responsibility, psychiatrists in the District of Columbia, as in New Hampshire before them, took it upon themselves to work out a legal concept of mental disease, first excluding, then including, sociopathic or psychopathic personality disturbances.

Eight years after *Durham* was decided the Court of Appeals came to recognize that "what psychiatrists may consider a mental disease or defect for clinical purposes, where their concern is for treatment, may or may not be the same as mental disease or defect for the jury's purpose in determining criminal responsibility".[8] The court therefore laid down a legal definition of mental disease for the first time:

A mental disease or defect includes any abnormal condition of the mind which substantially affects mental or emotional processes and substantially impairs behavior controls (*McDonald v. United States*, 312 F2d 851 (D.C. Cir., 1962)).

The Court of Appeals thus finally undertook a task it should have undertaken originally: to give legal meaning to the phrase "mental disease or defect" as it occurred in the legal rule of responsibility. Unfortunately, however, the definition adopted is simply a regression to the more traditional types of definition of legal insanity. 'Mental disease' is not itself actually defined, except insofar as a vague synonym is supplied: "abnormal condition of the mind." The informative part of the definition, which qualifies "abnormal condition of the mind," is simply a reversion to those ancient moral paradigms already incorporated in the M'Naghten test, the irresistible impulse test, and the American Law Institute test. Instead of "lack of knowledge" of the nature and quality of the act or its wrongfulness (M'Naghten) or instead of the "substantial capacity to appreciate the criminality of his conduct" (ALI), we have "substantially affects mental or emotional processes"; instead of language about acting under an irresistible impulse, or lacking "substantial capacity to conform his conduct to the requirements of law" (ALI), we have "substantially impairs behavior controls". Instead of really giving a legal definition of mental illness, the Court of Appeals in *McDonald* abandoned the attempt by following the traditional formula for legal insanity: mental illness, or some vague synonym for it, is only one element; added to it is some other traditionally excusing condition.[9] In doing so, the Court of Appeals abandoned the essentially correct insight behind *Durham* itself: that there is something about mental illness itself that excuses, irrespective of any ignorance about the nature of the particular crime or its prohibited nature, and irrespective of any issue of compulsion.

2. *The 'Albatross' of Durham: That Mental Illness Be the Cause of Crime*

In 1972 the Court of Appeals for the District of Columbia abandoned the *Durham* rule entirely and adopted the American Law Institute's definition of legal insanity (*United States v. Brawner*, 471 F2d 969 (D.C. Cir., 1972)).

One of the principal reasons for doing so was the problem that Court had been having with the 'product' portion of the *Durham* rule: psychiatric witnesses came to substitute their own judgments of the responsibility of the accused for that of the jury, and to phrase their conclusions on that ultimate issue in terms of whether or not the criminal act was the product of disease.

Contrary to some commentary on *Durham,*[10] this problem did not stem from the inherent lack of meaning that one can assign to the idea of mental illness causing crime. The concept of causation adopted by the court was unproblematic, and consistent with the analysis of the causal relationship in both many areas of law and much of contemporary philosophy.[11] The concept of mental illness as the sort of thing that could properly be said to be a cause of behavior was more problematic; for Judge Bazelon, the author of *Durham,* clearly did not believe that mental illness was necessarily some brain condition that caused behavior in some mechanical way.[12] Presumably he and other members of the court meant to invoke *psychological* theories of causation (as opposed to physical) in thinking of the mental conditions of the mentally ill as a cause of their behavior. The court certainly invited expert testimony in terms of such psychological theories:

> Description and explanation of the origin, development, and manifestations of the alleged disease are the chief functions of the expert witness . . . The law wants from the medical experts . . . expert medical opinion as to the relationship, if any, between the disease and the act of which the prisoner is accused. (*Carter v. United States,* 252 F2d 608, 617, (D.C. Cir., 1957))

The court got more than it wanted of such expert medical opinion. By 1967 it became clear to the court that psychiatric conclusions about 'product' were often disguised moral judgments about the culpability of the accused, so the court flatly prohibited psychiatric testimony in terms of 'product' or cause and effect (*Washington v. United States,* 370 F2d 444 (D.C. Cir., 1967)). It never became clear to the court why psychiatrists were substituting their moral judgments for those of the jury other than some speculations about defense counsel's strategies. For, if causation were a straightforward scientific question, why were the relevant scientists unable to testify to it without infecting their scientific judgments with their moral judgments?

The initial answer lay in the fact that causation was not an issue on which psychiatrists could be at all helpful, given the deterministic assumptions of modern psychiatry. For if psychiatrists, particularly those of a dynamic persuasion, took the causation test literally, then in every case in which a mentally ill person committed a crime they would have to find the crime to be caused by the disease. One accustomed to thinking in terms of the unified personality, of basic instinctual drives underlying all conscious motivations, of the pervasiveness of unconscious influence, and of the displaceability of psychic energy amongst seemingly unconnected objects, could reach no other conclusion. Accordingly, any distinction psychiatric witnesses might make between mentally ill defendants would have to be made covertly on non-causal grounds.

What psychiatry essentially lacked – and still lacks – was any reconciliation of its own deterministic assumptions with the concept of responsibility. If mentally ill persons are excused because of their lack of freedom in a contra-causal sense (i.e., freedom that more normal people enjoy), then psychiatry could be of no help for it has no concept of freedom the mentally ill are supposed to lack. Judge Bazelon himself ultimately came to wonder "how medical experts can be expected to provide information about the impairment of free will, when free will would seem to be a philosophical and not a medical concept."[13]

The ultimate reason for the psychiatric unhelpfulness was that causal connection in the sense used by the court was not the issue anyway. Mental illness is not an excuse from responsibility because it causes criminal actions. This is true no matter if 'mental illness' is construed as the name of some subset of physical causes themselves a part of a general mechanistic account of human behavior – the nineteenth century view – or if 'mental illness' is thought to name some subset of psychological causes themselves a part of a general 'para-mechanistic' account of human behavior – a twentieth century view. In either case, the theories involved would excuse us all if such causation excuses the mentally ill. Psychiatric theories about causation – this time in their much less precise twentieth century form of 'psychic determinism' – could accordingly only be a distraction from the actual criteria by virtue of which we separate the non-responsible from the responsible in law and morals.

Had the court paid more attention to the meaning of 'mental illness' and ignored any psychiatric theories of causation, it might have been able to have developed a meaningful idea of the relationship between the illness and the act. As is argued later in their paper, mental illness for moral and legal purposes means an incapacity for rational action. One may do a certain act in a certain way *because* of such incapacity, but the 'because' should not be construed on the model of mechanical or para-mechanical causation.

A bridge may collapse *because* its materials lack the tensile strength to hold it up; a person may fail a test *because* he is stupid. In neither case have we cited an event, contiguous in space and time to another event, such that the first can be said to cause the second.[14] The bridge falling or the person failing the test are events symptomatic of the general dispositional properties cited to explain them. Similarly, the relationship between illness and act the court might have sought was that the criminal act was symptomatic of the general incapacity for rational action of the mentally ill defendant. Because of its failure to work out an adequate legal definition of mental illness, however, the court could not reach this result but was bound to the psychiatric paradigm of mental illness that it had adopted, that of a cause of behavior.

IV. THE LEGAL PURPOSES BY VIRTUE OF WHICH MENTAL ILLNESS MUST BE DEFINED

The lesson of New Hampshire and *Durham* is plain: the law must give a legal definition of 'mental illness' in light of the purposes behind the rule of which the phrase is a part. To build an adequate definition of mental illness for appropriation in criminal law thus requires that one have clearly in mind the purposes behind having a defense of insanity at all. A rationale for such an excuse from punishment in turn depends on a clear apprehension of why we punish: the reasons for excusing a class of offenders from punishment necessarily depends on the reasons we have for punishing to begin with.

While this is not the place to work out a philosophy of punishment, one can perhaps define the issues. The central issue in the philosophy of punishment has been the question of the appropriate role to assign to the moral culpability of the offender. Two general positions may be distinguished: First, that moral culpability is irrelevant to punishability. We punish on this view in order to prevent crime. Punishment accomplishes this by deterring others, by incapacitating the criminal (via incarceration) from committing further crimes, and/or by reforming him to make him a safe citizen in the future. Second, that moral culpability is alone the only reason for punishing. Whether we punish, and to what extent, are on this view determined solely by the degree of moral culpability of the actor. Even if there is no need for deterrence of others, and even if the actor is not dangerous or in need of reform, he ought on this view to be punished solely because he deserves it.

The first of these views is generally thought of as utilitarian, and the second, as retributivist. There is a mixed view, which has gained considerable respect amongst contemporary philosophers and lawyers, which is that we require both that there be good utilitarian reasons to punish an individual, and that he be morally culpable, before punishment may be inflicted. Neither set of reasons is alone sufficient; both, that is, are necessary.[15]

Utilitarianism is inadequate as a theory of punishment for the simple reason that it sanctions unjust results. I will assume without argument here that we punish only because of the conjunction of good utilitarian reasons and the moral desert of the offender. This being so, there are two potential sets of reasons for having an excuse from responsibility such as insanity: that some class of offenders is undeterrable and non-dangerous, the utilitarian theory of excuses,[16] or that for one reason or another the class of offenders is not morally culpable. If both utilitarian and moral reasons are each necessary to punish, that means the absence of either is sufficient to justify an excuse from punishment; that is, that one ought to be legally excused either if he is not blameworthy or if his punishment would not serve utilitarian ends.

It is the lack of moral culpability, rather clearly, that has justified the defense of insanity since its inception in Anglo-American law. As stated by Bazelon in the *Durham* decision itself (*supra,* at 876), the legal and moral traditions of the Western world have long required that those who are mentally ill are not morally blameworthy, and not being morally blameworthy, cannot be the proper subjects of punishment. The moral nature of the justification for the defense is shown most markedly in those numerous jurisdictions with mandatory commitment statutes, for such statutes incarcerate those found not guilty by reason of insanity for the utilitarian reasons of incapacitation and reform. Such statutes make clear that for those mentally ill persons who are dangerous, the utilitarian reasons to punish are satisfied; it is only the lack of any moral culpability of the offender that requires that there be an insanity defense. The defense then functions solely as a morality play, segregating those who will be detained but who cannot fairly be made subject to the moral sanction associated with punishment.

The proper legal definition of mental illness, then, should reflect not medical classifications, nor necessarily utilitarian calculations about the dangerousness, deterrability or reformability of various types of mentally disturbed defendants. If the issue is a moral one, as it is on the mixed view of punishment (and as it would have to be on the retributivist view), then the legal definition of the phrase should embody those moral principles that underlie the intuitive judgment that mentally sick people are not responsible.

The legal definition of mental illness should thus draw on the moral tradition that is the rationale for the defense. What is thus needed is an analysis of that popular, moral notion of mental illness. What have people meant by the phrase 'mental illness', such that, both on and off juries, they have for centuries excused the otherwise wrongful acts of mentally ill persons? It is to this question that the balance of the paper is addressed.

V. THE ORDINARY CONCEPTION OF MENTAL ILLNESS

A natural way of proceeding into an analysis of the ordinary meaning of 'mental illness' would seem to be to analyze the meaning of each of its conjuncts: 'mind' and 'illness'.

A. The Meaning of Mind

One's idea of what constitutes a sick mind can be no better than one's idea of what a mind itself might be. Sir James Stephen, a noted nineteenth century criminal lawyer, perceived this:

> What is the meaning of the word mind? What is a sane and what is an insane mind? . . . Difficult and remote from law as some of these inquiries may be, it is impossible to deal with the subject at all without entering to some extent upon each of them ([56], pp. 128–129).

In answering the question "What is a mind?" in its most obvious sense, one runs immediately into a centuries-old philosophical thicket; that is, the ontological status of mind (and of all of the 'furniture' of the mind, such as intentions, pain, perceptions, etc.). The tempting modern positions in this debate seem to be four in number: (1) The dualist tradition of Descartes and Freud which asserts that apart from the physical world there exists the mental world, not existing in space but only in time, accessible only by the reflexive of consciousness. Mental phenomena, accordingly, are irreducible to physical phenomena, and an independent science of the mind, with an elaborate deductive structure mirroring the theoretical structure of physical science, is thus a legitimate enterprise. (2) The physicalist tradition, which in its modern form asserts that mind and brain are identical. There is just one species of existence, that of the physical world, on this view, and the ultimate reference of mental words would be found to be physical things ([3, 53]). (3) The behaviorist tradition, which interprets all reference to mind as constructs out of behavior [49, 51, 52]. In such a view mind is in one sense no thing at all, in the same sense that force is no thing at all for 'instrumentalists' discussing theoretical entities in the philosophy of science. Mind and its furniture, such as intentions, sensations, beliefs, etc., are only explanatory constructs, a collection of dispositions to behave in certain sorts of ways which explain behavior in the way the dispositional properties explain instances of the disposition in question for natural objects. (4) The most recent variant, functionalism, which asserts that minds are logical or functional states of physical systems [43, 44]. The analogy often drawn by proponents of this view is to a Turing machine, a machine whose states are defined by a machine table such that physical structure is left unspecified; that is, the machine may be said to be in a certain state by meeting criteria that have nothing to do with the machine's physical configuration. Mental states, analogously, are functional states of a person the criteria of which are intimately connected to his linguistic abilities, but whose physical realization is unspecified and perhaps unspecifiable in any one-to-one correspondence with his functional states.

Each of these views is not, as its counterpart in psychology might be, a 'theoretical perspective' or research strategy. Rather, each purports to tell us what we all ordinarily think, as presupposed in our ordinary speech, a mind is. They are, in other words, analyses of our shared conceptual system; not theoretical postulates in some new conceptual system. Thus, if correct, each of them would be the first step in analysis of our ordinary concept of mental illness.

In light of the well-charted difficulties with each of these positions, some way of avoiding a general defense of one of them would seem desirable.[17] One avenue of avoidance is that of 'semantic ascent': if one does not wish to talk about the thing itself, one may still talk about our talk about the thing. Indeed, much of the philosophy of mind prior to the emphasis of the last fifteen years or so on the mind/body relationship, is engaged in this kind of 'conceptual analysis': showing not what a sick mind is in terms of its relationship to physical bodies, but by analyzing 'mental illness' in terms of its logical relationships to other mental terms. One may leave the question of the reference of all mental terms open, and still say something about the internal relationships which hold between them.

In examining our mentalistic vocabulary, it is doubtlessly not necessary to fix the precise boundary which we use to mark off such vocabulary from the rest of our language. This is a matter of some dispute in philosophy, and for our purposes a precise demarcation is not in any event necessary.[18] Paradigmatically mental are words such as 'belief', 'knowing', 'motive', 'intention', 'imagining', 'perceiving', 'angry', 'depressed', and the like.

One set of characteristics of mentalistic expressions that some philosophers believe to be "the mark of the mental" has to do with certain peculiarities of our first person usage of many mental terms. Although often lumped together, there are in fact three distinct claims usually being made here. First of all, most of our vocabulary of mind is asymmetrical with respect to the modes of its verification in its first and third person usages. Contrary to the claims of behaviorists, I do not find out about my moods, emotions, motives, pains, imaginings, or whatever, in the same way that other people do; I do not, for example, observe myself saying 'ouch', or observe my hand being removed from the hot stove, in order to conclude that I am in pain. I know that I am in pain in an immediate way without resort to such observation, and my statement stating that I am in pain is thus verified in a way quite different from a similar statement made by someone else about me in the third person, namely, "he is in pain". Our mentalistic vocabulary, in other words, reflects a seeming 'privileged access' that we each have to our own mental states, privileged in the sense that first person claims to knowledge about mental states are not subject to the usual demands for evidence from observation.

Secondly, our language of mind is such that the words we use to describe many mental states make them seem to be self-intimating; if, for example, I am in pain, I know it. It is linguistically odd to ask oneself whether one is in pain. Thirdly, our mentalistic vocabulary seems to imply that we are incorrigible in our statements about being in mental states. If I think I am in pain, I am; if I think I intend to do '*x*', I necessarily intend to do '*x*'. We have, in other words, a seeming infallibility in our knowledge of our mental states reflected in the incorrigibility of our first person, present tense descriptions of them.

These three related characteristics reflected in the vocabulary of mind – privileged access, self-intimation and incorrigibility – apply with varying force to the different sorts of mental words. They do not, as is sometimes thought, all apply to our use of all mental words. Nonetheless, they are a pervasive feature of our mentalistic vocabulary.

None of these three characteristics apply to our usage of the phrase 'mental illness'. Mentally ill persons do not have some privileged access in determining whether they are mentally ill: if one is going to describe himself as being mentally ill, he is going to have to make the same sorts of observations as a third person observer would have to make, and does not have some non-inferential, non-observational way of knowing that he is mentally ill. The inference involved is shown directly by such common expressions as, "I must be insane" (compare: "I must be in pain"). Similarly, the state of being mentally ill is not self-intimating. One who is mentally ill need not necessarily know that he is mentally ill; often mentally ill persons go to great lengths to resist that conclusion. Finally, one is not incorrigible with respect to thinking that one is mentally ill. One can believe oneself to be mentally ill, and not be mentally ill at all. A psychiatrist need not accept at face value a patient's claim that he is mentally ill.

In this respect, 'mental illness' is like another class of mental words; namely, those words with which we describe the character traits of normal people, words such as 'greediness', 'vanity', or 'stupidity'. For there is also no privileged access, self-intimation or incorrigibility in our first person use of character-trait words either. We must observe our own actions and their motives to see if we are, for example, vain or greedy; we can easily be wrong if we think we are vain, and ignorant of the fact that we are greedy.

This parallel suggests that 'mental illness', like the words we use to describe character traits, is fundamentally connected to behavior and not to mental experience. For words of character are behavioral in their criteria, and this is reflected directly in the fact that first person usage of such character words is not incorrigible, nor is there privileged access, nor are such states self-intimating.[19] 'Mental illness' sharing those same features would seem to indicate a similar lack of connection to mental experience, but rather a connection to behavioral criteria in an application of the term. This is not to deny that being mentally ill may not involve characteristic mental experiences; only that those experiences are not themselves the criteria of the word. There may well be mental experiences characteristic of being greedy or vain as well; yet it is greedy and vain behavior, not those experiences, that entitle both the actor and the observer to so describe one's character.

To say that mental illness is connected to behavior is not yet to say what it is about the behavior of persons we label mentally ill, that enables us to so label them. We must defer this inquiry until we have first ascertained something of the ordinary meaning of the word 'ill', to which I now turn. Preliminarily, however, this 'third person aspect' of 'mental illness' may reflect the deep-seated 'we/they' attitude inherent in the popular understanding of the mentally ill,[20] because perhaps 'mentally ill' is a label we have invented principally to apply to other people who are different from us and not to apply to ourselves.

B. The Ordinary Meaning of Being Sick, Ill or Diseased

1. Illness and the Body

A popular misconception about our shared concept of illness reflected in the writings of psychiatrists like Thomas Szasz, is that to be ill one's bodily structure must deviate from a physiological or anatomical norm ([60], p. 23). There are, I suspect, two distinguishable claims here: (1) that illness is necessarily caused by something physical – if not some invading thing, like a spirochete in general paresis or a virus, then some physiological deviation of the subject's body, such as chemical imbalance in the neurotransmitters of the brain; and (2) that illness is necessarily manifested by deviant physiological structures, such as high blood pressure, inflamed joints, or whatever. There must be, in other words, physical symptoms and/or physical causes.

Neither of these claims squares with the ordinary way in which we use the word 'ill'. It is easy to see that such deviant physiological structures cannot be a sufficient criterion of being ill, else people with large heads, long noses, or abnormally structured bodily parts of any sort would all be ill. Yet many deviant structures are simply irrelevant to health or illness, and indeed some of them may make a person healthier.[21] More pertinently, such deviant physiological structures are not even a necessary criterion for being ill. Historically this had to be true; people were said to be ill long before anyone knew about the deviations of bodily structure that may accompany illness. Even today, when ordinary usage has had time to accommodate itself to the influence of medicine, we would describe someone as ill even if his bodily structure were completely non-deviant from that of the rest of the population. Imagine a gland universally present at birth in all human bodies that incapacitates the person from many activities and causes a great deal of discomfort. The fact that his body is statistically normal would not prevent us from saying that the individual is ill (indeed, wisdom teeth or appendices may approach this hypothetical gland).

None of this is as true of the word 'disease' as it is of 'ill' and 'sick'. For many doctors, "the term 'disease' is traditionally identified with pathology of tissue."[22] This medical assumption is reflected in ordinary usage, in two ways: first, we more often speak of a person ***having*** a disease, something which he possesses, than of him *being* diseased; secondly, the normal subject of the 'diseased' predicate is not the persons but their bodies or some part of their bodies. I shall pursue each nuance of ordinary usage separately below.

We quite readily say of someone that he has a disease; we do not as readily say he has a sickness or he has an illness. We are more likely in the latter case to say he is sick or he is ill. This reflects the fact that the concept of disease is linked to the classificatory systems of medicine more than are the concepts of being sick or ill. To say of someone he has a disease implies that he has some one disease amongst others. Saying that someone is ill or is sick in general does not presuppose he has some one sickness or some one illness amongst a host of classified illnesses and sicknesses.

The importance of this link of the concept of disease to systems of classification in the present context is this: classificatory systems are built up on implicit causal hypotheses, and the notion of disease reflects this origin. It sometimes is thought that diseases are classified by the recurrence of a pattern of symptoms alone:

> The term ['disease'] refers to a pattern of factors which somehow hang together and recur, more or less the same, in successive individuals. Thus, pain in the right lower quadrant of the abdomen, with nausea, vomiting, fever, and a high white count, spell out the features of acute appendicitis ([31], p. 197).

Yet this kind of aggregation of symptoms into a recurring pattern is only the first step in classifying diseases. Such correlations are made with the hope of finding a 'hidden nature' to such diseases, a hidden nature not to be found in the cluster of symptoms themselves. Once we discover such hidden natures (e.g., a ruptured appendix), the presence or absence of such determines whether a given individual has that disease; an individual who manifests identical symptoms without having a ruptured appendix does not have appendicitis. Moreover, the intention with which we speak when we use words such as 'polio' or 'multiple sclerosis' is to refer to such hidden nature even if we don't know it. The point, in other words, is not that we change the meaning of disease words when we discover physical causes, but rather that we intend by using such words to refer to some such set of causes, leaving to science to discover their actual nature.[23]

We use the word 'disease', then, in light of a set of background beliefs about there being discrete diseases, separate from one another not only by their differing cluster of symptoms but ultimately by their physical causes. There is thus at least a suggestion in ordinary usage that 'having a disease' means that something is wrong with one's body.

The second nuance of ordinary usage of 'disease' is that we freely predicate 'is diseased' to bodies and parts of bodies. We do this more frequently, indeed, than we speak of persons being diseased. We may say that his brain is diseased, or that he has brain disease, or that his body is "rife with disease". With 'ill' or 'sick', this is not the case: we do not usually say of someone that his body is sick or that his liver is ill – as if *he* were not. The general predicates 'is ill' or 'is sick' have as their normal subjects persons. 'Disease' in ordinary understanding is thus prima facie an attribute of bodies directly;

one would thus expect the symptoms and causes of the disease to be of a physical sort.[24]

For each of these reasons,[25] 'disease' does suggest physical symptoms and physical causes. Does this implication of physical causation 'infect' our use of the general predicates, 'is ill' or 'is sick'? It might seem that one cannot be ill without having a disease, that is, without having something wrong with one's body. Yet in fact the assumption of physical causation is not contagious. In the law of torts it was once correctly said that "there is no requirement whatsoever that a tort have a name", meaning that any personal injury caused by the culpability of another ought to be a compensable tort irrespective of whether or not it happened to fit the existing classification of causes of action that tort law has set up. A similar slogan regarding diseases would be appropriate: one can be ill without having any recognized disease. One is ill by virtue of being in the state characterized in the next subsection, a state principally characterized by incapacitation, and in some cases connected to pain and death as well. Only if one were to equate the general state of being ill with the more specific one of having some medically recognized disease, would the implication of physical causality be proper.[26]

Being ill, then, is independent of having some specific disease of the body, or of physical causation and symptomatology in general. In light of this, the common opposition of mental and physical illness can be misleading insofar as it suggests that real illnesses are physical illnesses, and that mental or functional illnesses are illnesses only by a rather weak analogy. Rather, people are just plain ill; if we wish to subclassify by causes or symptoms, all well and good. Only, one should not mistake the principle by virtue of which we make such subclassifications, with the general criteria for being ill to begin with.

Imagine a state, call it state A, defined by one criterion: being in the water 200 miles from the center of some island. Suppose, further, that this is thought to be an undesirable state to be in so that a profession is begun whose principal responsibility is to prevent people from getting into such a state. We find a number of persons who are in state A, that is, they are out in the water somewhere on the circumference of a circle whose center is the island, and whose radius is 200 miles. We would doubtlessly be curious about how they all got out there; indeed, to the professional group dedicated to preventing it, this knowledge of causation is crucial. We discover some got there by shipwreck; for the rest, we can discover no means of transportation (perhaps we do not know about airplanes or submarines). We may classify all such persons as being in "Shipwreck State A", and "Other State A". Both such groups of persons, however, are plainly in state A. How they got there is a very interesting question around which we may build a classificatory system; but it is also quite a distinct question from whether or not they are there, the sole defining characteristic of being in state A.

2. On Being Ill

If 'being ill' predicated of a person does not mean that that person's body deviates from some physiological norms, what does it mean? Under what conditions do we say of someone that they are ill or sick? Again, it should be recalled that we are seeking the ordinary meaning of 'ill'; not the kind of 'medical' meaning some physicians and sociologists of medicine have proposed, *viz.*, that 'illness' is a label whose use is governed by the appropriateness of medical treatment.[27] There is no such connection of treatment and illness in ordinary understanding: doctors treat matters that are not illnesses (e.g., delivery of babies, cosmetic surgery, circumcision, and matters of family planning); there are perhaps illnesses that doctors cannot treat: if it should turn out that mental illness is not physically caused, physicians limited to drug or other traditional medical therapies would not be the persons to treat them, although persons in such condition could still quite properly be considered ill. In addition, even if it should turn out that all illnesses are treated by doctors and that the only things doctors treat are illnesses, as a criterion for 'ill' it is about as helpful as "being the subject of concern by a mortician" is as a criterion for being dead. In both cases we should like to know the criterion by virtue of which doctors and morticians determine when it is appropriate to express their concern (i.e., when people are ill or dead).

The central idea behind illness is that of impairment: to be ill is to be impaired from functioning in some of the wide varieties of ways we think to be normal. Often such impairment may be accompanied by pain or other forms of distress; sometimes such impairment may be accompanied by an increased likelihood of death (which I suppose could be viewed as the limiting case of impairment of functioning). Each of these has been suggested as a supplementary criterion for being considered ill. But the central criterion of being ill seems to be that the person so described is incapacitated in some ways yet to be specified.

Functional impairment can occur at either of two levels: at the level of specific diseases of bodily parts; or at the level of the total functioning of the person in his society. Ultimately, the first depends on the second; that is, we judge someone to have a bad heart only by virtue of the heart not being able to perform its function properly, and we can judge whether or not the heart is properly performing its function only by knowing whether or not the person is or is not functioning properly.[28]

To see this, imagine someone who has a heart with a valve such that the tissues in various parts of his body receive an inadequate supply of blood so that he is seriously incapacitated; we will say of such a person that his heart is not performing its function properly. For the function of the heart is to circulate the blood, which by hypothesis it is not doing.

How we ascertain what the function of the heart is, and once having ascertained that, how we further ascertain whether it is adequately performing that function, illustrates the dependence of the notion of a properly functioning heart upon the notion of a properly functioning (i.e., healthy) person. The information content of function statements is that the part or process to which a function is assigned (the heart's beating) is a contingently necessary condition of the activity identified as the function (the circulation of blood).[29] Yet the assignment of functions to parts of a system cannot simply be made on the basis of this sort of causal information. Simply because the circulation of the blood is one consequence of the heart's activities, does not entitle us to label that consequence as the heart's function. For the heart's beating has a number of consequences in addition to the circulation of blood – for one thing, it makes some noise. Absent some other criterion of function assignment, we are equally free to say that the function of the heart is to make noise in the chest cavity.

The criterion by virtue of which we select out one consequence of the heart's activity as its function is that we already have some further end state in mind toward which that consequence itself contributes. Because the circulation of blood is itself necessary to the maintenance of the end state of the person properly functioning, it rather than the noise is assigned as the function of the heart.

Similarly with the determination of whether the heart is properly performing its function: the amount of blood that it is the heart's function to circulate is whatever amount is necessary for a human being to function properly.

To ascertain whether someone is ill, then, ultimately involves us in ascertaining whether he is incapacitated from functioning properly; even if his specific complaint involves some bodily defect, we will judge that defect to be a disease, and he to be ill, only if *he* is not in good working order.

How do we judge whether someone is in good working order? More specifically, what sorts of things must one be capable of doing in order to be not ill (i.e., healthy)? Many people are incapable of distinguishing good wines from mediocre ones; others seem to be incapable of experiencing orgasm in heterosexual relationships; others are incapable, because poor, from going to expensive artistic performances and are incapable of appreciating them if they were to go; still others are left-handed or stutter. Yet surely not all of these incapacities make one ill.

Our ordinary conception of health, or of proper functioning, avoids the pitfalls of equating health with happiness or with human flourishing in general, and of thus equating illness with any form of incapacity for happiness and flourishing, such as those involved in poverty, ignorance, and cultural deprivation. When we say of someone that he is ill, we mean he is incapacitated from pursuing those basic activities that are a part of (virtually) anyone's conception of a good life. In terms of our heart patient, he is ill if he cannot move about the world as well as most people without keeling over – in short, if he cannot do some wide variety of things we all do in our daily lives in order to get by. If he can do these basic things, he is healthy, that is, he is not ill.[30]

Ultimately of course what sorts of capacities are necessary for anyone's conception of the good life is culturally relative. We have good agreement that the ability to control one's body in the world is necessary; probably equally good agreement about the ability to think and reason coherently. Other societies could conceivably be different. In a society in which physical movements requiring any exertion were thought undesirable, health would not require the same sort of capacities as it does in our world. Our hypothetical heart patient would not be ill, nor his heart functioning improperly, just because he could not run up stairs, play football, etc. (although it is hard to imagine that he would not be incapacitated from some sort of activities thought to be desirable). All judgments of illness and improper functioning of body parts are thus in this sense normative; that is, dependent upon a society's agreed judgments about what sorts of basic capacities are desirable for any human being.

This hardly makes the ordinary conception of illness useless. We do have good agreement on certain basic capacities being essential to the good life, however conceived, of any human being in our society; it is that agreement which determines the boundaries of our ordinary conceptions of health and illness. It is only by leaving the ordinary conception of health, by expanding it to include all capacities one might think desirable in order to lead a fulfilled or happy life with a corresponding expansion of illness as anything less than such a life, that some psychiatrists and followers of naturalistic ethics make the normative aspects of health and illness troublesome.

There are a number of supplementary stipulations that one must make in order to narrow the extension of 'impaired basic capacities' to be equivalent to that of 'ill'. To begin with, we do not ordinarily think that permanent disabilities are illnesses. Persons who are born blind or deaf are not ill. We distinguish congenital conditions, no matter how seriously disabling, from illnesses. Secondly, we seem to distinguish injuries from illness too. The victim of an automobile wreck, no matter how incapacitated, is not in ordinary understanding ill, although he obviously is not healthy either. These distinctions are reflected in the criminal law's distinction between disease and defect:

We use 'disease' in the sense of a condition which is considered capable of either improving or deteriorating. We use 'defect' in the sense of a condition which is not considered capable of either improving or deteriorating and which may be either congenital, or the result of injury, or the residual effect of a physical or mental disease. (*Durham v. U.S.*, *supra*, at 875)

There are doubtlessly other limitations as well that must be pursued in a thorough ordinary-language analysis of the meaning of 'ill'. One would need to exclude some conditions capable of improving which do involve basic incapacities, such as pregnancy and perhaps extreme obesity.[31] More crucial, however, would seem to be the claim that the foregoing analysis of 'ill', far from being too inclusive, is too narrow. For recurrent in the legal, philosophical, and medical literature on the nature of illness is the suggestion that pain (or more broadly, distress) and death are independently sufficient criteria of being ill [31, 58, 63]. Pain or distress, however, while often accompanying illness, and often motivating the seeking of medical treatment, is neither necessary nor sufficient as a criterion of illness. One may suffer real pain, yet if one is not also incapacitated by the pain one is not ill; one can be paralyzed, and thus ill, if it is a condition capable of improving, without the presence of any pain.

Death as a criterion is somewhat more plausible. The argument often advanced [30] in favor of treating death as itself a sufficient criterion stems from what we say of the man who drops dead suddenly without a day's incapacitation or a single painful sensation. If he died from a slowly growing cancer in the lung, but had no pain or incapacitation, was he ill before he died? We would clearly say he was diseased, or that he had a lung disease, or that his body was rife with disease. But that's because of the tie of 'disease' to pathology of tissue noted earlier. 'Sick' is more ambiguous: we typically say of such cases *both* that "he was never sick a day in his life", and "he was a very sick man but didn't know it". 'Ill', however, is not applicable to such persons: we would not say "he was ill for a long time before he died". He was not ill because he was not in any way disabled.[32] Hence, death, while it doubtlessly is a criterion by virtue of which we judge deviant physical structure to be disease,[33] is not a criterion of being ill.

3. *Illness and Responsibility*

There is one further apparent limitation to our concept of illness, which seems to have direct bearing on responsibility assessments in law and morals, and thus deserves some discussion. The idea is suggested in the ancient notion of a patient as one who is passive. Becoming ill, it might be thought, is necessarily something that happens to one, not something one brings on oneself. Since we are responsible only for what we do, not for what happens to us, illness necessarily excuses us from responsibility for its symptoms, on this view of the meaning of 'illness'.

Some such view is implicit in Samuel Butler's satire on punishing the ill for the symptoms of their illness in *Erewhon*. It is explicit in one recent philosophical account of 'illness' ([9], pp. 61–62). It was accepted by at least some of the members of the U.S. Supreme Court in their decision on whether drug addicts and alcoholics are ill and thus exempt from punishment under the cruel and unusual punishment clause of the federal Constitution. To begin with the latter, the majority opinion in *Robinson v. California*, 370 U.S. 660 (1962), assumed for largely unstated reasons that any condition properly thought of as a disease could not be punished:

> It is unlikely that any State at this moment in history would attempt to make it a criminal offense for a person to be mentally ill, or a leper, or to be afflicted with a venereal disease; . . . a law which made a criminal offense of such disease would doubtlessly be universally thought to be an infliction of cruel and unusual punishment . . . Even one day in prison would be a cruel and unusual punishment for the 'crime' of having a common cold (*ibid*, at 666–667).

It was not until *Powell v. Texas*, 392 U.S. 514 (1968), that four of the members of the Court made clear what it was about a condition being a disease that had to do with its punishability. In *Powell*, the dissenters found that the 'core meaning' of the idea that alcoholism is a disease "is that alcoholism is caused and maintained by something other than the moral fault of the alcoholic, something that, to a greater or lesser extent depending upon the physiological or psychological makeup and history of the individual, cannot be controlled by him" (*ibid*, at 561).

There are in fact two limiting conditions on the meaning of 'illness' implied in the foregoing: (1) that if one by his voluntary act brings on the condition, he is not ill; and (2) if one can remove the condition by voluntary act, he is not ill. Neither of these conditions is part of the ordinary meaning of 'illness'. If they were, one who had lung cancer due to smog would be ill, but one who had the identical condition exclusively because of smoking would not.[34] Similarly, people who have curable diseases would not be thought to be ill if they continued in their state for voluntary failure to take the known cure; indeed, this would seem to have the curious result that Christian Scientists are never ill except when afflicted with incurable diseases.

While the central idea of illness is the impairment of function, the abilities that are impaired when one is said to be ill do not necessarily include the ability to get out of the condition one is in. That ability is possibly relevant to the medical profession determining whether drug addicts and alcoholics are worthy of medical attention; it is possibly relevant as well to the law's concern about the fairness of punishing one for being in a condition he cannot help; the only point here is that it is not denied simply because one is properly said to be ill.[35]

Hence, to those who argue that mental illness is a misnomer because of the patient *choosing* to be in his incapacitated state ([10], p. 171), one might reply that such choice, even if truly made by mental patients, does not preclude them from being properly considered 'ill'. Only if one could show that mental patients were in fact not at all incapacitated – e.g., that they could if they wanted to, disbelieve their hallucinations or delusions, could act rationally – would one be entitled to conclude that they were not ill. As I have argued elsewhere [35], this has not been shown.

C. On Conjoining 'Mind' and 'Ill': the Ordinary Meaning of 'Mental Illness'

What do we add to the meaning of 'ill' when we describe someone as being *mentally* ill? The first question to be resolved is whether 'mental' describes the symptoms of an illness, or its causes. It is often assumed by those hostile to the idea of mental illness that the classification is by causes. Thus, the arguments of Szasz are proffered against those psychiatrists who believe physical causation to underlie mental illness: if there are such physical causes, according to Szasz, then we have brain disease, not mental illness ([60], pp. 12–24).

It is odd to think that something of such long standing as our concept of mental illness could be the hostage of physiological research in this way; that is, that such research could show us, not more about mental illness, but that it really does not exist. If the hope of nineteenth-century psychiatry were realized so that every person that we call mentally ill possessed within his body some identifiable physical condition, surely we would say that we have discovered mental illness to have physical causes, not that it didn't exist.

This suggests, although it does not prove, that the classification of illnesses as mental is related in some way to the symptoms exhibited by the person, not to the species of causation involved. What sort of symptoms lead us to classify an ill person as being *mentally* ill? It is that set of symptoms showing some impairment of those abilities we think of as mental abilities. Our mentalistic vocabulary is rich in words labeling, by some functional division, various mental powers. We have capacities of perception, of memory, of imagination, of learning, the basic capacities of reasoning and thinking, the capacities to feel emotion, and the capacities of will to have one's emotions and desires issue in one's actions. It is the impariment of these mental functions that we first and foremost have in mind when we speak of someone as being mentally ill. Thus, the ancient synonym of mental illness, as being a person who has "lost his reason".

Recalling the behavioral bias of our criteria in applying 'mental illness', it would seem that the kinds of symptoms we should look to that would exhibit in a relevant way the sorts of mental incapacities we have in mind when we call someone mentally ill, should be behavioral symptoms. Although psychologists or others might test mental capacities in ways quite different from simply observing daily behavior; and while it is doubtlessly a legitimate enterprise to attempt a phenomenological description of what it is like to experience going crazy; still, the principal criteria by virtue of which we ordinarily apply our notions of mental illness should be in some way linked to the behavior of the person whom we can call mentally ill.

Perhaps the analogy to character traits will again prove fruitful. When we say of another that he is greedy or vain, we have necessarily generalized over large aggregations of his observed behavior. We are claiming to have observed a consistency in his actions, namely, that he characteristically performs greedy or vain ones. Further, we are claiming in our description of him as greedy or vain that he is disposed to do similar sorts of things in the future. We thus both find an intelligible pattern to his past actions, and make a hypothetical assertion about what he will do if he gets the chance in the future.

'Mentally ill' is constructed somewhat similarly. We apply the label to a class of persons whose actions follow a certain pattern. Unlike words of character, however, we predicate 'mentally ill' of a person whenever we find his pattern of past behavior unintelligible in some fundamental way; or perhaps because of the unintelligibility of his acts, the only pattern we discern is that there is no pattern at all, and thus we think such an individual to be unpredictable as well.

When is there this fundamental failure of understanding of a fellow human being? To answer this requires that one be familiar with the kind of understanding that we each have of ourselves and others in daily life. Let me begin with two very ordinary examples.

Suppose we ask Smith why he is carrying his umbrella to lunch. He replies that he believes it is going to rain. Suppose we ask Jones why he went across the street a moment ago, and he replies it was because he wanted some tobacco from the store across the street. Each, by his answer, has rendered his behavior intelligible by giving us the practical reasoning that led him to do what he did. Each has supplied a part of the practical syllogism of his action, to use Aristotle's terminology.

For Aristotle, the full practical syllogism had two premises: (1) specifying what the agent desired, and (2) the beliefs that he had as to the means available to that desire's fulfillment. The conclusion of the syllogism was the action itself. In ordinary speech we usually need only one part of the syllogism to make out the other. For example, Smith specified the belief while Jones specified the desire. The more complete practical syllogism for Smith would be: (1) I do not want to get wet on my walk to lunch; (2) I believe that it is going to rain and unless protected by an umbrella I will get wet on my walk to lunch; therefore, I will take my umbrella. Jones's practical syllogism, in terms of his (stated) desire and his (unstated) beliefs, is equally obvious.

These examples bring out a simple fact: We explain the actions of our fellow men by reference to their beliefs and their desires (the latter includes motives, wants, wishes, intentions, aims, goals, objects, purposes, etc.). By citing such beliefs and such desires, we see the action as a rational activity of an agent who is using the action to achieve something he wants. Such explanations rationalize an action in that they exhibit it as the rational thing to do given certain beliefs and certain desires. For example, if Smith were not rational, he could have a desire to stay dry, a belief that taking his umbrella was necessary for him to stay dry, and yet deliberately leave his umbrella in his office.

We may schematize the practical syllogism so as to make explicit the presupposition of rationality implicit in Aristotle's version. The full syllogism in general form should have four premises: (1) Agent *X* wants result *R* to obtain; (2) *X* believes that in situation *S* action *A* will cause *R* to obtain, and that he is in situation *S;* (3) if *X* believes that *A* will result in *R,* and if *X* desires *R,* then *ceteris paribus, X* will do *A;* (4) *ceteris paribus.* Therefore, *X* does *A.* In other words, we need to know: (1) what the agent wanted, (2) what he believed about the situation and his ability to achieve through action what he wanted. In addition, we need to know (3) that he is a rational creature (that is, other things being equal, one who will act so as to further his desires in light of his beliefs), and (4) that the agent does not have desires and beliefs that conflict with the desires and beliefs on which he is about to act.

The unintelligibility of the actions of the mentally ill stems from the fact that this form of explanation is not as regularly available to explain their actions as it is to explain the actions of the rest of us. More specifically, this form of explanation may break down in any of four ways, corresponding to each of the four premises of the practical syllogism.

In extreme cases, we may be unable to make out any set of beliefs and desires by virtue of which we may view an action as rational. For example, the bodily motions of epileptics during a grand mal seizure, or perhaps the word salads of schizophrenics would seem to be nonrational activities in this sense. In such cases we are unable to see the action as the rational thing to do in light of any set of beliefs and desires that we can reasonably ascribe to the agent. Second, to say that a person's action is irrational may be to deny the rationality of the beliefs on which the action is based (premise 2). The individual who avoids all contact with people because he believes he is made of glass and will shatter if touched behaves rationally enough in light of his desire not to be shattered and his belief that he is made of glass and will shatter if touched. But the belief itself is irrational, and actions predicated on irrational beliefs are themselves, in ordinary understanding, irrational actions.[36]

Third, we may find the agent's desires defective because it is not intelligible to us that one could want what the agent apparently wants. It is evident that not all consequences of an action render that action intelligible if they are cited as the actor's motives. To be a motive for action, a consequence must in addition be intelligible as an object of desire in a given culture. For example, if we come across an individual carrying books up the stairs of his house and ask why he is doing so, he may reply that he simply wants to put all the green books in his library on the roof. He desires this, we may suppose, for no further reason. As it stands, such a desire is unintelligible to us as a motive for action, even though given such a desire the man's act in fulfilling it is rational enough.[37]

Finally, the fourth premise is rarely satisfied for the mentally ill. Their desires are often inconsistent: by acting to fulfill one desire, they often frustrate another. While this circumstance is true for most of us some of the time, more of the mentally ill's desires are unconscious and, being unconscious, cannot be resolved into a coherent, consistent set of wants. Acting on such unresolved conflicts of desire cannot be fully rational, because such actions necessarily frustrate some things one wants a great deal, although perhaps unconsciously.[38]

When explanations of the actions of another human being break down in these ways, we say that he is irrational. We mean by 'irrational' just that some significant portion of his actions pursues ends that are unintelligible to us, are predicated on beliefs themselves irrational, are based on gross inconsistencies of desires or of beliefs, or perhaps are not based on desire or belief sets at all.

It is this kind of pattern of irrational action that is the primary symptom of the mental incapacities we label mental illness.[39] For the mental abilities of perception, memory, imagination, and particularly reasoning, are necessary in the acquisition of rational beliefs, and the rendering of consistency between belief sets and between desire sets. Rationality is the fundamental premise by virtue of which we understand ourselves as human beings; that is, as creatures capable of adjusting their actions as reasonably efficient means to intelligible ends. Being mentally ill means being incapacitated from acting rationally in this fundamental sense.

There are obviously degrees of irrationality. How irrational must one be to be mentally ill in popular understanding? Psychiatry in this century has doubtlessly influenced the popular understanding of the concept of mental illness. We are all to some degree irrational in our conduct; we are thus tempted to say that we are all a little bit crazy.

Yet side by side with this sophisticated, educated view there exists our ancient paradigm of mental illness: it is not manifested by the occasional irrationality we all exhibit. It is reserved for those gross deviations from intelligibility we still capture with the more severe statements, 'he is crazy', or 'he is insane', or 'he is mad'. Those idioms capture the essential notion of the ancient conception of mental illness as madness: that mentally ill people are different from us in ways that we find hard to understand.

The National Opinion Research Center at the University of Chicago, in surveying the public's ideas about mental illness, found "that there is an old, socially-sanctioned, well-established set of views which supports the identification of mental illness only with the violent, extreme psychosis and, within this context of ideas, mental illness emerges as the ultimate catastrophy that can happen to a human being".[40] In this set of views, the Center further concluded, severely diminished rationality and lack of moral responsibility were prominent criteria. It is, as I shall next argue, this set of views about severe irrationality that the law should adopt and which, in practice if not in theory, has in fact long been adopted in utilizing the concept of mental illness for legal purposes.

VI. THE PROPER LEGAL DEFINITION OF MENTAL ILLNESS IN THE CONTEXT OF LEGAL INSANITY

What is it about the popular understanding of mental illness that precludes moral responsibility and thus precludes punishability by the criminal law? Why, in other words, does severely diminished rationality excuse? It is because our notions of who is eligible to be held morally responsible depend on our ability to make out rather regularly practical syllogisms for actions. One is a moral agent only if one is a rational agent in the sense of 'rational' that I have discussed. Only if we can see another being as one who acts to achieve some intelligible end in light of some rational beliefs will we understand him in the same fundamental way that we understand ourselves and our fellow men in everyday life. We regard as moral agents only those beings we can understand in this way.

Thus, societies that attribute moral responsibility to natural objects are regarded as primitive, because we are unwilling to accept the primitive animism subscribed to by such societies. If we did believe that natural objects possess beliefs and desires in light of which their movements are to be understood as rational, our misgivings about attributing moral responsibility to such objects would evaporate. Moral responsibility and that mode of explanation I have called practical reasoning go hand in hand.

It is easy to understand the longstanding historical tendency of the criminal law to analogize the mentally ill to infants and to animals. For we do not think of these beings as engaging in practical reasoning to the same extent as do normal, adult human beings. Only when an infant begins to act in ways explicable by his practical syllogisms do we begin to see him as a moral agent who can justly be held responsible. The same is true of the mentally ill.

The proper definition of legal insanity is one which utilizes this moral criterion. If criminal law is to reflect our shared notions of culpability, an excuse from punishment based on those moral notions ought to utilize those same moral criteria. The only question appropriate to juries is thus one appealing to their moral paradigm of mental illness: is the accused so irrational as to be non-responsible? The other definitions of legal insanity all mistakenly transport basic excuses from elsewhere in morals and criminal law, to the mentally ill to whom they have only a partial application.

One rather suspects that juries have long applied this criterion irrespective of the wording of the insanity test in which they are instructed by the judge:

> However much you charge a jury as to the M'Naghten Rules or any other test, the question they would put to themselves when they retire is – "Is this man mad or not?" [48]

The reason why they have done so is because they, along with those many lawyers and psychiatrists who have thought 'insanity' and 'psychosis' to be roughly equivalent,[41] have perceived that madness itself precludes responsibility.

University of Southern California
Law Center, Los Angeles, California

NOTES

[1] In Bk. III, Ch. 1 of the *Nicomachean Ethics* Aristotle subdivides actions that may be regarded as involuntary (and thus excused) into two classes, those done in ignorance and those done under compulsion.

"Virtue or excellence is, as we have seen, concerned with emotions and actions. When these are voluntary, we receive praise and blame; when involuntary, we are pardoned and sometimes even pitied. Therefore, it is, I dare say, indispensible for a student of virtue to differentiate between voluntary and involuntary actions. . . .

. . . It is of course generally recognized that actions done under constraint or due to ignorance are involuntary. An act is done under constraint when the initiative or source of motion comes from without. It is the kind of act in which the agent or the person acted upon contributes nothing. . . .

. . . Ignorance in moral choice does not make an act involuntary – it makes it wicked; nor does ignorance of the universal, for that invites reproach; rather it is ignorance of the particulars which constitute the circumstances and the issues involved in the action. . . .

. . . As Ignorance is possible with regard to all these factors which constitute an action, a man who acts in ignorance of any one of them is considered as acting involuntarily, especially if he is ignorant of the most important factors. The most important factors are the thing or person affected by the action and the result."

Aristotle: 1962, *Nichomachean Ethics*. Martin Ostwald, trans. Bobbs-Merrill Co., Inc., New York, pp. 52–57.

[2] I am indebted to Joel Feinberg [18] for seeing this general point.

[3] Two influential expressions of this view were Lord Erskine's eloquent defense in

Hadfield's Case, 27 How. St. Tr. 1281, 1314 (1800), in which he asserted that "delusion ... where there is no frenzy or raving madness, is the true character of insanity", and Sir John Nicholl's opinion in *Dew v. Clark*, 3 Add. Eccl. Rep. 79, in which Nicholl said: "The true criterion, the true test of the absence or presence of insanity, I take to be the absence or presence of what, used in a certain sense of it, is comprisable in a single term, namely, 'delusion' ... I look upon delusion in this sense of it, and insanity, to be almost, if not altogether, convertible terms."

4 In *Commonwealth v. Rogers*, 7 Metc. 500 (Mass. 1844) Ray's views were presented at the trial by Ray himself; among a number of other tests, Chief Justice Shaw, presiding at the trial, told the jury that they were to find "whether the prisoner in committing the homicide acted from an irresistible and uncontrollable impulse", paraphrasing rather closely Ray's earlier language that some mentally ill persons are "irresistibly impelled to the commission of criminal acts". See Ray ([45], Section 187).

5 *State v. Pike*, 49 N.H. 399, 442 (1869). This notion of mental disease as physical cause made the irresponsibility of the mentally ill self-evident to many other nineteenth-century judges as well. See, for example, *Commonwealth v. Mosler*, 4 Pa. St. 264 (1846), in which Chief Justice Gibson spoke of "an unseen ligament pressing on the mind, drawing it to consequences which it sees but cannot avoid ...".

6 This notion – that causation precludes agency – is found in some of the work of contemporary philosophers as well. See, for example, R. Taylor [62].

7 The 'verdict' of the State Hospital psychiatrists in New Hampshire has in practice been accepted as authoritative regarding the mental illness of any individual accused. See J. P. Reid [46].

8 See *McDonald v. United States*, 312 F2d 847, 851 (D.C. Cir., 1962). In explaining the *McDonald* decision some years afterwards, Judge Bazelon noted that "... by casting the test wholly in terms of mental illness [in *Durham*] we had unwittingly turned the question of responsibility over to the psychiatric profession ... [In McDonald] we made 'mental illness' a legal term of art for the purposes of the insanity defense and told juries and judges that they could find mental illness when the psychiatrists found none ..." [5].

9 Judge Bazelon himself came to recognize this. In his opinion in *United States v. Brawner*, 471 F2d 919, 1011 (D.C. Cir. 1972), he noted that "The definition of mental disease adopted in McDonald rendered our test, in almost every significant respect, identical to the ALI test."

10 Thomas Szasz attacked the product requirement as "unadulterated nonsense" [59].

11 In *Carter v. United States*, 252 F2d 608 (D.C. Cir., 1957), the court analyzed the causal relationship to be that of *sine qua non* familiar throughout the law: "But for the disease the act would not have been committed." On the use of this concept throughout the law, see H. L. A. Hart and A. M. Honore [26].

12 One knows this because in *Stewart v. United States*, 214 F2d 879 (D.C. Cir., 1954), decided only two weeks after *Durham*, Judge Bazelon reversed a trial court instruction that had equated mental illness with brain disease.

13 See *United States v. Brawner*, 471 F2d at 1029 (D.C. Cir., 1954). Cf. Alan Stone [57]: "Psychiatry failed Judge Bazelon because like all of the social and behavioral sciences it lacks a concept of an evil-meaning mind as a cause ... "

14 'Mechanical cause' as I have been using the phrase has four requirements: (1) that *events*, not states, be cited as the cause; (2) that such events be connected with the event

they are given to explain by a covering law connecting such classes of events, in a form supporting a counter-factual inference; (3) that such events be contiguous in space and time; and (4) that the relationship be temporarily asymmetrical. E. Nagel ([38], p. 73) attempts to isolate a common conception of cause citing similar factors. By 'para-mechanical cause' I mean that the same stipulations are made, except that the events and their contiguity are not spatial but mental or psychological; see G. Ryle [49].

15 Leading statements of this view of punishment will be found in H. L. A. Hart [25] and in Herbert Packer [39].

16 The classic example of this theory of excuses is to be found in Chapter XIII of J. S. Bentham [7].

17 One apparent avenue that might seem tempting is to discern that the proper subject of the predicates 'is ill' and 'is sick' is not either mind or bodies, but persons; that accordingly, one need not ask what sorts of things minds are to understand the concept of mental illness, because a sick mind is not involved, only a sick person. While I believe 'is ill' to be a predicate whose proper subject is persons, the conclusion allowing one to avoid the sticky question of the ontological status of mental entities, does not follow; for unless one abandons entirely the notion of mental illness, and speaks only of illness as such, one will still need to mark off those illnesses that are mental from those that are not. Presumably this would be done either in terms of mental symptoms or mental causes, or both. In any case, one will inexplicably be bound up in saying something about the nature of mental causes, as opposed to physical ones, or symptoms manifested 'in the mind' as opposed to physiology. The 'person-is-primitive' notion fails to avoid the mind/body problem, for understanding mental illness as well as more general problems.

18 The line of demarcation is often attempted in terms of a characteristic much emphasized by Brentano, that of 'intentionality'. See [13], pp. 19–42.

19 Thus, the repeated use of character words as examples in a behavioral view of mind such as in Ryle [49]. 'Mental illness' was also used by Ryle as a seemingly obvious example of the correctness of the behaviorist view. See ([49], p. 21): " . . . for all that we can tell, [on a non-behaviorist account] the inner lives of persons who are classed as idiots or lunatics are as rational as those of anyone else. Perhaps only their overt behavior is disappointing . . . ."

20 See S. A. Star [55], a research study conducted by the National Opinion Research Center of the University of Chicago in the midfifties, in which a significant percentage of the thirty-five hundred interviewees thought both present and former mental patients to be quite different and unpredictable; reprinted in part in Donnelly *et al.* ([16], pp. 818–820).

21 See L. S. King [31]: ". . . ordinarily statistics alone cannot label any part of the data as 'diseased'. When we apply statistical methods we already have in mind the idea of health." See also the excellent discussion of this point in S. J. Morse: 1978, 'Crazy Behavior, Morals and Science: An Analysis of Mental Health Law', *Southern California Law Review* **51**, 527–654, at 569.

22 See Hinsie and Campbell [28], defining 'psychosis' as a disorder but not a disease.

23 See generally, H. Putnam [41] and [42], wherein Putnam rejects the analysis of the meaning of disease words as being simply a cluster of symptoms, even if the causes are unknown.

24 Although again, this does not mean that physical deviation is a sufficient criterion of having a heart disease any more than it is of being ill in general. Abnormal physical

structures of a particular organ will only constitute a disease of that organ if those structures impair the functioning of the organ.

25 Other distinctions between illness and disease not here germane would include the fact that 'ill' is predicated principally of persons whereas 'diseased' will be predicated of any living system. We readily talk of plants being diseased, of there being plant diseases; we do not talk of a plant being ill. Using 'ill' also seems to involve us more often in acts of evaluation than use of 'disease', suggested by usages such as being in 'ill-health' and as the proverbial ill wind that blows no one any good, when 'ill' is used simultaneously with 'poor' and 'bad'. On these distinctions, see Christopher Boorse [9].

26 Psychiatrists have had an unfortunate tendency to regard as mental illness any condition listed as a mental disorder in *DSM-II* (*Diagnostic and Statistical Manual of Mental Disorders,* 2nd ed., American Psychiatric Association). This would make our concept of illness (here, mental illness) dependent upon our empirical success in isolating syndromes and their causes. Our ideas about who is ill or sick are not dependent upon particular schemes of classification, as many psychiatrists have come to recognize. See Busse [11].

27 See E. Jellinek ([29], pp. 58–59), concluding that alcoholism is an illness if the medical profession recognizes it as such. See also F. Kraupl Taylor [61], equating being diseased with patienthood and the latter with expression of therapeutic concern by doctors.

28 Compare Louis H. Swarz ([58], p. 404) where it is suggested that to hold the line between ethics and medicine disease (equated with illness) be limited to less than total human functioning.

29 Much of what follows draws heavily on C. Hempel [27]. See also J. L. Cohen [12] and E. Nagel ([38], pp. 401–428, 520–535).

30 I have assumed in the text that 'health' is best analyzed as meaning only 'not ill'. There is doubtlessly a positive sense of 'health' as well, connected with feelings of vitality and a vigorous sense of wellbeing (e.g., as in "I feel so healthy today"). See G. H. Von Wright ([63], pp. 50–62), in which the 'privative' and 'positive' senses of 'health' are distinguished.

31 These and many other apparent counter-examples are discussed in Robert L. Spitzer and Jean Endicott's article [54]. See also my article [36] in the same book.

32 Compare Kendall ([30], p. 5): "A man with a cancer growing silently in his lung, or someone with anginal pain which he dismisses as a touch of wind, would both be regarded by both doctors and laymen as ill . . ." Since Kendall does not distinguish illness from disease, the disagreement is perhaps not as great as it seems.

33 This is why we speak with such ease of plant disease – not because plants are incapacitated or suffer, but because they are in a state correlated with an early death. We do not speak of a plant *being ill,* however. See note 25, *supra.*

34 Similarly, we will say of someone that he is mentally ill even if we believe that he got himself into that state by the voluntary taking of drugs. See *People v. Kelley,* 111 Cal. Rptr., 17p, 516 P2d 875 (1973), in which the defendant was thought to be psychotic despite the court's belief that her condition was due to the voluntary ingestion of drugs.

35 Cf. H. Fingarette [20], where Fingarette argues that punishability ought not to be settled by the decision whether or not to call alcoholism a disease.

36 On the notion of rational belief, see R. J. Ackermann [1].

37 The example is from G. E. M. Anscombe [2]. For further analysis of intelligibility of ends, see A. J. Watt [65]. Intelligibility of ends should be distinguished from irrationality of ends. When we say that an end pursued by another is irrational, we usually mean that it does not comport with what we think he needs. Such judgments about what a person needs are usually thinly disguised normative judgments about what a person ought to want. If one knows the rationality of ends a human being should follow, such judgments might be justified; one might then be entitled to expand the notion of health along the lines some psychiatrists have proposed, to encompass all human flourishing, and illness, correspondingly, as anything less than that. Our shared concept of mental illness does not reflect such knowledge, however. We do not regard people as mentally ill because their ends do not agree with our own; it is only if their ends are so far removed from our own experience that we cannot understand how anyone could want such a thing – if in other words their ends are unintelligible – that we judge them as mentally ill.

38 See Harvey Mullane [37], who characterizes the neurotic as "irrational because what he does is regularly and systematically self-defeating".

39 This analysis of mental illness as lack of rationality is derived from my article [34]. Other analyses of mental illness as irrationality are H. Fingarette [21, 22], Feinberg [18], and Mullane [37].

40 See Star [55]; also see R. Waelder ([64], p. 384), who notes a 'common core' to the concept of mental illness centered on "conditions in which the sense of reality is crudely impaired, and inaccessible to the corrective influence of experience – for example, when people are confused or disoriented or suffer from hallucinations or delusions. That is the case in organic psychoses, in schizophrenia, in manicdepressive psychosis. Their characterization as diseases of the mind is not open to reasonable doubt."

41 See Goldstein ([23], pp. 59–62), where he reports a "quite widespread feeling among psychiatrists that all psychotics should be regarded as insane . . .". See Livermore and Meehl ([32], p. 804): "the criminal law has usually used the term [mental disease] in what one writer has called its core concept . . . " (quoting Waelder [64]).

BIBLIOGRAPHY

1. Ackermann, R. J.: 1972, *Belief and Knowledge*, Doubleday, Garden City, New York.
2. Anscombe, G. E. M.: 1957, *Intention*, Cornell University Press, Ithaca.
3. Armstrong, D.: 1968, *A Materialist Theory of the Mind*, Routledge and Kegan Paul, London.
4. Baier, K.: 1970, 'Responsibility and Action', in M. Brand (ed.), *The Nature of Human Action*, Scott Foresman and Co., Glenview, Illinois, pp. 100–116.
5. Bazelon, D. L.: 1971, 'New Gods for Old: "Efficient" Courts in a Democratic Society', *New York University Law Review* 46, 658–659.
6. Becker, L. E.: 1973, 'Durham Revisited', *Psychiatric Annals* 3, 8.
7. Bentham, J.: 1789, 'Cases Unmeet for Punishment', in *An Introduction to the Principles of Morals and Legislation* (1948: Hafner Press, New York).
8. Biggs, J.: 1955, *The Guilty Mind*, Harcourt, Brace, New York.

9. Boorse, C.: 1975, 'On the Distinction between Disease and Illness', *Philosophy and Public Affairs* 5, 49–68.
10. Braginsky, B. M., Braginsky, D.D. and Ring, K.: 1969, *Methods of Madness: The Mental Hospital as a Last Resort,* Holt, Rinehart, and Winston, New York.
11. Busse, E. W.: 1972, 'The Presidential Address: There Are Decisions to be Made', *American Journal of Psychiatry* **129**, 1–9.
12. Cohen, J. L.: 1955, 'Teleological Explanations', *Proceedings of the Aristotelian Society* **51**, 255–292.
13. Dennett, D. C.: 1969, *Content and Consciousness,* Humanities Press, New York.
14. Dennett, D. C.: 1973, 'Mechanism and Responsibility', in T. Honderich (ed.), *Essays on Freedom and Action,* Routledge and Kegan Paul, London, pp. 157–185.
15. Diamond, B. L.: 1956, 'Isaac Ray and the Trial of Daniel M'Naghten', *American Journal of Psychiatry* **112**, 651–656.
16. Donnelly, R. C., Goldstein, J., and Schwartz, R. D.: 1962, *Criminal Law,* Free Press of Glencoe, New York.
17. Epstein, I. (ed.): 1935, *Babylonian Talmud, Baba Kamma,* Soncino Press, London.
18. Feinberg, J.: 1970, 'What Is So Special About Mental Illness?', in *Doing and Deserving,* Princeton University Press, Princeton, Chapter 11, pp. 272–292.
19. Fingarette, H.: 1967, *On Responsibility,* Basic Books, New York.
20. Fingarette, H.: 1970, 'The Perils of *Powell*: In Search of a Factual Foundation for the "Disease Concept of Alcoholism" ', *Harvard Law Review* **83**, 793–812.
21. Fingarette, H.: 1972, 'Insanity and Responsibility', *Inquiry* **15**, 6–29.
22. Fingarette, H.: 1972, *The Meaning of Criminal Insanity,* University of California Press, Berkeley.
23. Goldstein, A.: 1967, *The Insanity Defense,* Yale University Press, New Haven.
24. Group for the Advancement of Psychiatry: 1954, 'Criminal Responsibility and Psychiatric Expert Testimony', G.A.P. Report No. 26.
25. Hart, H. L. A.: 1968, *Punishment and Responsibility,* Clarendon Press, Oxford.
26. Hart, H. L. A. and Honore, A. M.: 1959, *Causation in the Law,* Oxford University Press, New York.
27. Hempel, C.: 1965, 'The Logic of Functional Analysis', in *Aspects of Scientific Explanation,* Free Press, New York, pp. 297–330.
28. Hinsie, L. E. and Campbell, R. J.: 1960, *Psychiatric Dictionary,* 3rd ed., Oxford University Press, New York.
29. Jellinek, E.: 1960, *The Disease Concept of Alcoholism,* Rutgers Center of Alcohol Studies Publications, New Brunswick, New Jersey.
30. Kendall, R. E.: 1974, 'The Concept of Disease and its Implications for Psychiatry', Inaugural Lecture, No. 57, University of Edinburgh; reprinted in 1975 in *British Journal of Psychiatry* **127**, 305–315.
31. King, L. S.: 1954, 'What Is Disease?', *Philosophy of Science* **21**, 193–203.
32. Livermore, J. M. and Meehl, P. E.: 1967, 'The Virtues of M'Naghten', *Minnesota Law Review* **51**, 789–856.
33. *Model Penal Code,* § 4.01 (Proposed Official Draft, 1962).
34. Moore, M. S.: 1975, 'Mental Illness and Responsibility', *Bulletin of the Menninger Clinic* **39**, 308–328.
35. Moore, M. S.: 1975, 'Some Myths About "Mental Illness" ', *Inquiry* **18**, 223–265.

36. Moore, M. S.: 1978, 'Definition of Mental Disorder', in Robert L. Spitzer and Donald Klein (eds.), *Critical Issues in Psychiatric Diagnosis*, Raven Press, New York, pp. 85–104.
37. Mullane, H.: 1971, 'Psychoanalytic Explanation and Rationality', *Journal of Philosophy* 68, 413–426.
38. Nagel, E.: 1961, *Structure of Science: Problems in the Logic of Scientific Explanation*, Harcourt, Brace, and Jovanovich, New York.
39. Packer, H.: 1968, *The Limits of the Criminal Sanction*, Stanford University Press, Stanford, California.
40. Platt, A. and Diamond, B. L.: 1966, 'The Origins of the "Right and Wrong" Test of Criminal Responsibility and Its Subsequent Development in the United States, an Historical Survey', *California Law Review* 54, 1227–1260.
41. Putnam, H.: 1962, 'Dreaming and Depth Grammar', in R. J. Butler (ed.), *Analytical Philosophy*, 1st series, Basil Blackwell, Oxford, pp. 36–48.
42. Putnam, H.: 1965, 'Brains and Behavior', in R. J. Butler (ed.), *Analytical Philosophy*, 2nd series, Basil Blackwell, Oxford, pp. 1–20.
43. Putnam, H.: 1966, 'Minds and Machines', in S. Hook (ed.), *Dimensions of Mind*, New York University Press, New York, pp. 148–179.
44. Putnam, H.: 1964, 'Robots: Machines or Artificially Created Life?', *Journal of Philosophy* 61, 668–691.
45. Ray, I.: 1838, *A Treatise on the Medical Jurisprudence of Insanity*, Little, Brown and Company, Boston.
46. Reid, J. P.: 1960, 'The Working of the New Hampshire Doctrine of Criminal Insanity', *University of Miami Law Review* 15, 14–58.
47. Reik, T.: 1953, 'The Doe-Ray Correspondence: A Pioneer Collaboration in the Jurisprudence of Mental Disease', *Yale Law Journal* 63, 183–196.
48. *Royal Commission on Capital Punishment 1959–53 Report*, London, 1953, § 322.
49. Ryle, G.: 1949, *The Concept of Mind*, Barnes and Noble, New York.
50. *Select Pleas of the Crown*, Volume I, Selden Society, London, p. 30 (1887).
51. Skinner, B. F.: 1953, *Science and Human Behavior*, Free Press, New York.
52. Skinner, B. F.: 1973, *More About Behaviorism*, Free Press, New York.
53. Smart, J. J. C.: 1959, 'Sensations and Brain Processes', *Philosophical Review* 68, 141–156.
54. Spitzer, R. L. and Endicott, J.: 1978, 'Medical and Mental Disorder: Proposed Definition and Criteria', in Robert L. Spitzer and Donald Klein (eds.), *Critical Issues in Psychiatric Diagnosis*, Raven Press, New York, pp. 15–39.
55. Star, S. A.: 1955, 'The Public's Idea About Mental Illness', paper presented to the Annual Meeting of the National Association for Mental Health, Sheraton-Lincoln Hotel, Indianapolis, Indiana.
56. Stephen, J.: 1883, *A History of the Criminal Law of England*, Vol. 2, Macmillan, London.
57. Stone, A.: 1975, *Mental Health and Law: A System in Transition*, Center for Crime and Delinquency, National Institute of Mental Health, Rockville, Md., p. 226.
58. Swarz, L. H.: 1963, 'Mental Disease, the Groundwork for Legal Analysis and Legislative Action', *University of Pennsylvania Law Review* 111, 389–420.
59. Szasz, T.: 1958, 'Psychiatry, Ethics and the Criminal Law', *Columbia Law Review* 58, 183–198.

60. Szasz, T. S.: 1970, *Ideology and Insanity,* Doubleday, Garden City, New York.
61. Taylor, F. K.: 1971, 'A Logical Analysis of the Medico-Psychological Concept of Disease', *Psychological Medicine* **1**, 356–365.
62. Taylor, R.: 1965, *Action and Purpose,* Prentice-Hall, Englewood Cliffs, New Jersey.
63. Von Wright, G. H.: 1963, *The Varieties of Goodness,* Humanities Press, New York.
64. Waelder, R.: 1952, 'Psychiatry and the Problem of Criminal Responsibility', *University of Pennsylvania Law Review* **101**, 378–390.
65. Watt, A. J.: 1972, 'The Intelligibility of Wants', *Mind* **81**, 553–561.
66. Weihofen, H.: 1960, 'The Definition of Mental Illness', *Ohio State Law Journal* **21**, 1–16.
67. *Year Books of Edward II,* Year Books Series, Vol. V, Selden Society, London (1909).

SECTION II

CRIMINAL AND CIVIL LIABILITY OF THE MENTALLY ILL

JEROME NEU

MINDS ON TRIAL[1]

> . . . even a dog distinguishes between being stumbled over and being kicked.
>
> *Holmes* ([15], p. 3)

> After everything is settled, Chico objects to one last provision. "That's in every contract", Groucho assures him. "That's what they call a sanity clause". "You can't fool me", laughs Chico. "There ain't no Santy Claus."
>
> *A Night at the Opera* [33]

Why is the criminal law concerned with the contents of people's minds, what sort of concern is it, and what is it concern for?

I

There have been times and places when having certain thoughts was itself a crime. A person's religious beliefs, for example, provided grounds for prosecution during certain periods of history. This is no longer so (at least in our criminal justice system). The change has not come about simply because religion no longer matters. Nor has it come about because of skepticism about the possibility of knowing other minds, of uncovering hidden or unexpressed beliefs – indeed, the new toleration generally allows for freedom of expression, so one may express as well as merely hold unpopular beliefs. It has become clear the purposes of the law can be served provided that people observe the requirements of the law in their behavior. For example, the purpose of social cohesion is served even if people have beliefs that might lead them to interfere with others, provided that they refrain from doing so. The beliefs in themselves are not dangerous. Second, beliefs are governed by rational constraints. It is unfair, if not ineffective, for the law to command belief where the evidence will not support

B. A. Brody and H. Tristram Engelhardt, Jr. (eds.), Mental Illness: Law and Public Policy. 73–105.

it. And where the evidence does support it, the law adds nothing to that support. It is irrational, if not impossible, to make ourselves believe against the evidence as we see it – for the law to require someone to try would be to command self-deception. We can, however, in general refrain from prohibited actions and conform to the outward requirements imposed by the law. Though our freedom of thought has been hard won, it depends in the end on essential differences between thought and action.

The closest our law comes to punishing for thoughts alone is perhaps in crimes of attempt. But even here there must be more than just thoughts. Intention alone is not enough, one must have taken steps in preparation for and towards completion of the crime (be it attempted robbery, or attempted murder, or anything else). And the outward steps are not simply *evidence* for certain desires or intentions.[2] They are an essential part of the offense. The mere having of a hostile wish is not a crime (even if the person tells us about it); it is acting on the wish, attempting to harm someone, that brings the machinery of the law to bear.

We do not punish for thoughts alone. The having of a certain state of mind, whether it be belief, desire, intention, or whatever, is not – as things stand – sufficient to make a person punishable by law. This is not to say that we do not find certain states of mind reprehensible. We may disapprove of or in any case wonder about a person who takes joy in the sufferings of a friend or who fails to feel grief at the death of a loved one – but we do not punish people for such inappropriate responses. We may view such responses as signs of bad character, but we do not punish people for character (or lack of it) alone.

II

Using the term 'thoughts' broadly – to cover a wide range of mental states, including belief, desire, intention, and so on – the mere having of certain thoughts is not sufficient for punishment under the law; it is, however, in general, and for good reason, necessary to have certain thoughts. Most offenses include, in addition to the doing of certain overt acts, a condition of *mens rea* – criminal intention, a guilty mind, or evil intent (literally: a 'criminal mind'). The psychological states are usually built into the very definition of the offense. The causing of the death of another does not constitute 'murder' unless it is done with *intent* to kill or at least to cause grievous bodily harm. In any case, the psychological states of *mens rea* come out in the conditions that are allowed as excuses. What might otherwise look like murder is not murder if it was done involuntarily, or accidentally, or in a dozen other ways that preclude the requisite intention, or, more interestingly for our purposes, if done by a person who lacks the requisite capacities.

The requirement of *mens rea* reflects the moral principle that a person is not to be blamed for what he has done if he could not help doing it. In our Western tradition 'ought' implies 'can', and a person who could not help doing what he did is not morally guilty – any more than an infant or a chair which causes harm is morally guilty. So liability to punishment is usually excluded if the law was broken unintentionally, under duress, or by a person judged to be below the age of responsibility or to be suffering from certain sorts of mental abnormality.[3]

The question now becomes how does one establish the requisite psychological conditions – i.e., *mens rea* (a guilty mind)? Given that an action was done, how does one establish whether it was done with or without the requisite intention? Here I can only provide a sense of the complexity of the issues and of their importance.

How are we to look into the mind of another? Certainly there is room for skepticism. A medieval English justice held: "The thought of man is not triable; the devil alone knoweth the thought of man"[4] – and others hold similar views today.[5] But certainly we can, sometimes at least, know what another person believes, desires, hopes, fears, etc. We can do this even in the case of seemingly private sensations. A person may know that he is in pain, directly and infallibly, by feeling it. But we may know it too. That we may know it in ways less certain than his, does not mean that we do not know the same thing. We too have access to his mind. We may know that he is in pain because he tells us. But we may know it, even when he denies it, through its manifestations in behavior. And even if he controls his behavior (as in the story of the Spartan youth whose belly was being gnawed at by an animal) we may know it, simply from his situation and from what we know about how human beings work.

In the law there is always the evidence of behavior. But it must be emphasized that it is only evidence. A too crude behaviorism in psychology is misleading, and in the law can be unjust. Behaviorism in psychology is the doctrine that psychological states, such as anger or belief, are reducible to, or in any case may be directly inferred from, observable patterns of behavior. Hence, being angry is supposed to be simply a matter of angry behavior or dispositions to such behavior. To see how misleading this sort of reduction can be, consider the following case:

Suppose you see *A* punch *B*. Further suppose that is all you see and all you know of their behavior and relation. Now, what emotion is *A* expressing? One is inclined to say 'anger' and perhaps 'jealousy' and other emotions in that range (i.e., unpleasant and hostile). But why not, say, 'gratitude'? Perhaps *A* is grateful to *C*, who hates *B*, and *A* expresses that gratitude by hitting *B*. Perhaps *A* is grateful to *B*, but *B* has strange ways of deriving pleasure (or at least *A* believes *B* derives pleasure in those strange ways). The point is that any bit of behavior can express (almost) any emotion [18]. Given an action, a bit of behavior, it is not yet determined what emotion (if any – a punch may be thrown on a bet) is being expressed. It is not just that there may be cultural variations; even within a culture (as the example shows) there are ambiguities and so problems of interpretation. More is needed than the isolated action – one must come to appreciate the context: not the actual context or the context as seen by you, but the context as understood by the agent; his beliefs, his desires, and his intentions; that is, his reasons for action. And you cannot read his intentions or reasons from the behavior and circumstances as seen by you or any other impartial observer.

It is sometimes claimed that the law has no interest in reasons or motives, that it is interested only in intentions. But it is difficult to see how one would describe (let alone account for) the role of 'provocation' or 'duress,' or excuses of too much or too little motive (e.g., kleptomania), or motives of the wrong sort, or even the nature of crimes of further intention (where something is done 'with the intent to' – so 'intent' has to do with the reason for the action and not just foresight of consequences), without reference to the character of motives or reasons or desires. And, in any case, intentions raise problems of interpretation of their own: behaviorism accounts as little for intentions as for the other psychological states mentioned.

III

So much – though it is admittedly far too little – for psychological behaviorism. A similar crude behaviorism sometimes prevails in the criminal law, as when the requirements of *mens rea* in the actual offender are allowed to lapse in favor of a standard provided by 'the reasonable man'. The standard provides what is sometimes called an 'objective' rather than a 'subjective' test of intention.

There is a presumption in law that a man intends the natural consequences of his acts. But what are those natural consequences? Certainly not everything that happens subsequent to them. Are they then the foreseeable consequences? If so – foreseeable to whom? To the jury, to the rational man whoever he may be (in England he is the man on the Clapham omnibus), to the accused, if to the accused, to the accused at the time of trial or at the time of the crime? The truth is that a person is often unaware of what might be called the 'natural' consequences of his acts, and that even where he might normally, i.e., usually, be aware of them, he might not have been on the particular occasion in question. Where this is not due to negligence or recklessness, but rather some lack of capacity or unusual circumstances, it seems fair to say that the person could not help it and so unfair to subject him to the full punishment of the law. Insanity provides the extreme case. Short of insanity, where one judges by the standard of the rational or reasonable man, one may be unfair to the actual man before one, ascribing to him an intention he did not and, in the circumstances and with his capacities, could not have. This occurred in the important case of Smith in England in 1961. As a result of the ensuing controversy, the law was changed.

The facts were as follows. Smith was driving a car containing stolen property when a policeman told him to draw into the kerb. Instead he accelerated and the constable clung to the side of the car. The car zig-zagged and collided with four oncoming cars; the policeman fell off in front of the fourth and was killed. Smith drove on for 200 yards, dumped the stolen property, and then returned. He was charged with murder and convicted by the jury ([16], p. 146).

It was never suggested that Smith intended to kill the policeman, but rather at the most, to cause him grievous bodily harm, and Smith himself contended that he "only wanted to shake him off". In any case, by asking what the reasonable man would have foreseen, rather than what Smith actually foresaw, the court may have been unfair to him. One cannot hold that if a reasonable man would have foreseen grievous bodily harm, that proves that Smith himself foresaw it. One might hold that if a reasonable man would have foreseen grievous bodily harm, then it is for the accused to prove that in a particular case he did not. But even so, "does a reasonable man ever get into a panic? If not, then what a reasonable man would have foreseen gives us little guidance as to what Smith, in a panic, foresaw. If so, then it seems that in order to arrive at a verdict in cases such as Smith's, the jury must know when it is, and when it is not, reasonable to get into a panic."[6] For these, and other reasons, the law in England now requires that juries apply a subjective and not an objective test in ascertaining the intent required in murder, using all the evidence and information available and not just the standard of a 'reasonable man'. In cases of diminished responsibility (as with Smith), as in cases of provocation, the result of a finding of absence of intention is not total excuse, but a lesser charge (manslaughter instead of murder) necessitating lesser punishment.

Now objective standards, such as the test of the reasonable man, need reflect neither skepticism about the law's ability to ascertain subjective facts nor a crude behaviorism. It may, as in the case of Justice Holmes, reflect a social policy, that is, a choice among competing values. Holmes's view was that: "Public policy sacrifices the individual to the general good ... For the most part, the purpose of the criminal law is only to induce external conformity to rule ... the tests of liability are external, and ... of general application. They do not merely require that every man should get as near as he can to the best conduct possible for him. They require him at his own peril to come up to a certain height. They take no account of incapacities, unless the weakness is so marked as to fall into well-known exceptions, such as infancy or madness. They assume that every man is as able as every other to behave as they command."[7] Holmes believed this was as it must and should be. Hence we see that objective standards may sacrifice respect for individuals and the concerns guarded by requirements of *mens rea* to what is regarded as social welfare. This happens even more clearly in those areas (happily very few) in which the law applies a standard of *strict liability:* given that a certain thing (e.g. selling adulterated food) has been done, you are responsible, there is no excuse. No test of mental state, objective or otherwise, is applied. The law simply probihits the doing of certain things and does not worry about intentions or states of mind at all. Strict liability, like taboo, allows no excuse. It does not allow for the possibility that the action could not have been helped, that the person was not in a position, or lacked the power, to avoid it. And it is for that reason objectionable.[8] It violates the "right of men to be left free and not punished for the good of others, unless they have broken the law when they had the capacity and a fair opportunity to conform to its requirements" ([14], p. 244). The requirements of conditions of *mens rea* are meant to safeguard that right.

IV

The defense of insanity opens up a trial to a full exploration of the mental states and capacities of an accused person. Before moving on to that exploration, its meaning and purposes, I would like to explore a few further complications in the notions of intention and action involved in the presumption in the law that a man intends the natural consequences of his acts.

First, in relation to 'intention', as we have seen, this presumption extends the ordinary notion of 'intention' in several different ways, depending on what one counts among the 'natural consequences' of a person's acts. But if they include only the actually foreseen consequences, or those foreseeable to the man at the time, I think the notion of intention is extended in a perfectly acceptable way. It is extended to include what are sometimes called 'oblique' or 'indirect' intentions: If you set off a bomb in a public place (as Desmond and Barrett did in 1868 in an attempt to free some Irish colleagues from an English prison by dynamiting the prison walls) though in fact it may have been your intention only to destroy the building and you had no interest in harming innocent passersby, the law nonetheless regards such harm as among the foreseeable and normally doubtless foreseen consequences of setting off such a bomb. And so, though you did not directly intend the harm, and may indeed regret it, you are nonetheless responsible and punishable for it. Such foreseen but perhaps undesired consequences are the objects of 'oblique intention' (the objects of 'direct intention' – the ordinary notion – are both foreseen and desired). The reason for regarding people as responsible for oblique intentions (that is, for the foreseeable natural consequences of their acts) is plain: given that they were foreseeable and that you nonetheless did the act, then, though they may not have been directly desired they must nonetheless have been acceptable to you, and it is the business of the law to make them unacceptable to you. As H. L. A. Hart, the distinguished English philosopher of law, puts it: direct and oblique intention can be "equally wicked, equally harmful, and equally in need of discouragement by the law" ([14], p. 127).

Now I would like to turn to some further complications, ingredient in the notion of action. These are complications that affect all attributions of intentions and responsibility. We have seen that there is a presumption in the law that a person intends the natural consequences of his acts. Let us ask again, what does this mean? If you pull the trigger of a gun, you cannot plead that you did not mean for the bullet to come out and kill, that you meant only to pull the trigger. One's actions cannot be so easily cut off from their consequences. But sometimes, in truth, one *is* only testing the trigger mechanism, and if a death results it *is* unintended, it is an accident. Whether the fact that the result was neither desired nor foreseen would in a particular case excuse depends on the circumstances. The lack of foresight in certain circumstances may itself constitute culpable negligence or involve recklessness.[9] But the point is that which consequences count as natural consequences of one's action, and so as part of one's intention, is not always clear. The line between the natural consequences and the irrelevant consequences of an act is not itself natural. But some such line must be drawn. One cannot be held responsible for *all* the consequences of one's actions. Some of them are extremely remote in time or place and unforeseen if not unforeseeable. Others may be 'consequences' in only a tenuous sense, one's action being only a minor contributory factor to the result. The matter depends partly on context and partly on purposes, but in any case it is essentially disputable. And the difficulty about distinguishing natural from other consequences of an action is really part of a more general difficulty about distinguishing an action from its consequences (was my action 'pulling a trigger' with the consequence that someone died, or was my action 'killing that person' with the consequences whatever they were, all being subsequent to his or her death). And this difficulty is, in turn, part of a yet more general difficulty about the specification of action: actions may be described, understood, and interpreted on many levels, and the level of conscious intention is only one among them, and not necessarily the most important.

Consider a man making marks with a pen. What is he doing? It might seem that there is a basic description of what the man is doing in terms of his bodily movements: he is moving his hand (or, more strictly, his hand moves) in certain ways. But think of the vast number of things he may be doing through the same bodily movements: perhaps he is idly making marks, but perhaps he is writing his name; he may be signing a check or a document. It might therefore be true to say, in answer to the question, "What is he doing?", that he is killing someone – as true perhaps as it would be to say it of the man who pulls a trigger. This last point illustrates that, on some levels and for some purposes, very different bodily movements may make for the same action. And a further intention may make the same bodily movement into a different action: consider *wounding* done with an intent to kill, as opposed to wounding – as by a surgeon – with intent to heal. Now what was the man *really* doing? The question is indeterminate. The relevant description of his action depends on one's interests and purposes. If one is being scientific, one might stick to a physical description level and say "he is moving his hand". One might remain objective and say, "he is writing". Or one might give an intentional description, and say "he is writing his name". But even here, is one to say he is doing only that which he explicitly has in mind to do? For there are unconscious purposes, and, as we have seen, oblique intentions. And whatever an individual may think he is doing, his action may have certain consequences, and even if he does not foresee them and they do not fulfill unconscious purposes, they may nonetheless be real and important: a check I sign may be cashable even if I thought I was only giving a handwriting sample at the time. How to differentiate the content of an action from its consequences? Where to draw the line? Perhaps from a moral point of view the intentional description is most important – what matters is what you think of yourself as doing. But again, consequences also matter, and also one may be wrong, or self-deceived. The complexities are endless. But the point is that when one talks of someone having done a certain thing and then tries to raise the question of responsibility as though it were a separate issue, one may fail to see that in describing his action in a certain way (in saying what the person did) you may already build in or leave out things essential for the determination of responsibility. The specification and attribution of an action is itself disputable. What the most appropriate description is of what a person did will depend on all sorts of things, many of which cannot be determined separately from settling issues of responsibility. One cannot pick out a person's 'behavior' in some neutral way and then say that whatever else may be true they (e.g. all who commit 'acts of violence') have a behavior problem calling for treatment or punishment. Again we see, from another direction, that one cannot read off the nature of an action or intention on the basis of 'objective', observable behavior alone. The law must inquire into mental conditions and capacities.

V

Returning to the place of *mens rea*, that is, the mental element, in the law. The law, in determining *mens rea*, usually makes the assumption that it is dealing with a reasonable man. Given the absence of the relevant excusing conditions (e.g., accident), a person is presumed to have the requisite capacities for *mens rea* and the law does not inquire further into his state of mind or capacities. It usually assumes that the psychological conditions of responsibility are met. "The insanity defense marks the transition from the adequate man the law demands to the inadequate man he may be" ([12], p. 18).

The traditional formulation of the relevant principles is provided by the M'Naghten rule, the language of which derives from a case in England in 1843. The rule tells jurors:

> that every man is to be presumed to be sane, . . . and that to establish a defence on the ground of insanity, it must be clearly proved that, at the time of the committing of the act, the party accused was labouring under such a defect of reason, from disease of the

mind, as not to know the nature and quality of the act he was doing; or if he did know it, that he did not know he was doing what was wrong. (*M'Naghten's Case,* 10 C.&F. 200, 210–211; 8 Eng. Rep. 718, 722–723 (1843).)

Now, it is a principle in the law that, in general, ignorance of the law is no excuse. But that, clearly, is not the sort of ignorance in question here. For one thing, the rule focuses on the particular act committed, and ignorance of circumstances ('mistake of fact') is a traditional defense. The judges go on to say that a person suffering from an insane delusion "... must be considered in the same situation as to responsibility as if the facts with respect to which the delusion exists were real". For another thing, please notice the M'Naghten rule is concerned with the *capacity* to know and to reason, and in particular with impairments in that capacity deriving from mental disease. The tests it lays down have been much criticized, and even, on occasion, modified. The Model Penal Code of the American Law Institute, for example, adds to the mental defect the knowledge conditions, a condition concerning capacity for self-control:

A person is not responsible for criminal conduct if at the time of such conduct as a result of mental disease or defect he lacks substantial capacity either to appreciate the criminality of his conduct or to conform his conduct to the requirements of law [1].

Note that this new rule maintains the reference to 'mental disease' that occurred in M'Naghten. The Durham rule of 1954 actually reduces the essential definition to mental disease itself as a condition of responsibility, leaving off specific tests (other than the 'product' requirement):

an accused is not criminally responsible if his unlawful act was the product of mental disease or mental defect. (*Durham v. U.S.*, 214 F.2d 862, 874–875 (D.C. Cir. 1954).)

Now, though the matter is important, I do not wish here to go into the niceties of the legal definition of insanity. What I want to emphasize is that it is a *legal* matter, that is, a matter involving social and moral values and principles, and not simply a medical-scientific matter.

At the heart of all the formulations concerning the insanity defense is the notion of 'mental disease' or 'mental illness'. And 'mental disease' is not a medical notion. It is not a diagnostic category within psychiatry. Of course, one *could* give it a psychiatric meaning. One could, for example, equate it with *all* illnesses treated by psychiatrists or, alternatively, with the psychoses. Let us consider these two suggestions. I wish to suggest that neither would serve any significant psychiatric purpose, nor would either of them capture the range of excusing conditions the insanity defense is meant to cover, or bring out the importance of the defense to our legal system.

The first proposal, that mental disease be equated with all illnesses treated by psychiatrists, immediately encounters two sorts of general skepticism. First, there is widespread skepticism about a disease model for mental disorders. It is often pointed out [25] that whether we regard certain character traits, desires, etc. as symptoms of mental disease depends on certain social and ethical norms. The problem was dramatically underscored when St. Elizabeth's Hospital in Washington, D.C., reversed an earlier position and decided that 'sociopath' was henceforth to count as a mental disease not merely a personality disturbance, thereby (under the then prevailing Durham rule) turning it into an exculpating condition. Thus, as Judge Burger put it, the Hospital's staff was able to "alter drastically the scope of a rule of law by a 'week-end' change in nomenclature which was without any scientific basis" – and a man who had been found guilty of murder only a month before had to be given a new trial.[10] Unexamined norms and social factors may determine which range of problems come before psychiatrists in the first place, and similarly unexamined norms may determine which among the problems they treat psychiatrists count as 'mental disease.'[11] What counts as and gets treated as disease, perhaps even physical disease, depends on what matters to you and your society. The notion of mental disease is not a neutral scientific notion, but depends on social norms. All sorts of problems end up in the hands of psychiatrists for all sorts of reasons. It cannot be left to them to decide which among the problems they deal with constitute legal insanity.

Now there is a second sort of general skepticism which puts more emphasis on problems with the concept 'mental' than with 'disease.' What is it that makes a disease 'mental?' At one time it was thought that all mental diseases are diseases of the brain, and this thought, that there is an underlying neurophysiological account of mental disorder, remains the hope of many (it was in fact the hope of Freud). But if there is an underlying neurophysiological explanation for each mental disorder, we do not know it yet. So we cannot differentiate mental from physical disorders by saying mental diseases involve malfunctioning or damage to the brain rather than some other bodily organ. Indeed, though much could be said about causes, and symptoms, and treatments, and about techniques and problems of differential diagnosis (say in cases of conversion hysteria), it is best to admit that there are enormous confusions and complications in nosology, the science of classification of diseases [22].

We frame the world in accordance with our interests and understanding, and in sorting out diseases and emotional disorders there are alternative competing and complementary rationales; we can divide them according to symptomatology, aetiological and developmental factors, prognoses, or the treatments to which they are amenable. These alternative schemes may overlap, or similar cases may fall under different names, if we change the basis for our nosological scheme. What is important is that we be aware of what we designate and on what grounds. Debates about the reality of 'schizophrenia', for example, may be understood as disputes about whether it denotes a natural kind or an odd conglomeration of symptoms. Is there a common aetiology (e.g., a biochemical abnormality or a problem in one's psychological development during childhood) and so a common treatment (or at least a 'family resemblance' among schizophrenic aetiologies and treatments)? And different conclusions may be based on different predictions about the development of science. But here I do not wish to press nosological skepticism any more than I pressed the skepticism based on "the myth of mental illness". The point is that equating mental disease with all problems treated by psychiatrists is of no help to psychiatrists with their problems of causation, and treatment, and classification, and it will not help us to understand why 'insanity' should be a defense against the imputation of responsibility in the law. Not every nervous tic or neurosis constitutes an excusing condition. Nor does a general condition excuse in a particular case without some specific connection being shown. Nor does the existence of an alternative treatment by itself mean there should be no liability to punishment: people might stop stealing if the state gave them money, so it does not *have to* imprison them. Indeed, what counts as an 'alternative treatment' may be becoming unclear: "We have . . . reached the position in which an important branch of medical practice consists simply of talk, even if this is a rather special form of talk . . . . When this stage is reached, however – when the medical profession has enlarged its toolbox to make room for words alongside of its traditional bottles, drugs and forceps – at this stage the definition of 'medical' methods becomes infinitely elastic: and there is no longer any logical reason why the medical treatment of crime should not be interpreted as covering all known methods of dealing with anti-social persons" ([30], p. 243). The conditions of 'rehabilitation' (what is needed to prevent an offender from recidivism) may not properly determine culpability for the offense in the first place.

I mentioned an alternative way of giving a *medical* sense to the notion of mental disease referred to in M'Naghten and the other insanity defense rules: it could be taken to cover just the psychoses. But then the notion would be redundant. It would be of no help to the psychiatrist or no more help than just the notion of 'psychoses'. And though the notion of 'mental disease' might then carry the implication, like psychosis, of some sort of loss of contact with reality (which is the main distinguishing feature between psychoses and neuroses), it would still remain unclear what the connection is between mental disease and criminal responsibility. The mere having of a disease, mental or physical, is not by itself excuse for anything and everything – even if it is severe and arouses sympathy. There must be some specific connection between the having of the disease and the doing of the particular (otherwise) punishable act. Psychoses, where they include delusions, may seem to match the cognitive test of the M'Naghten standard for criminal insanity. But not all psychoses are characterized by delusions (e.g., manic-depression) and there are non-psychotic forms of cognitive impairment (e.g., gross defects or retardation). More generally, other conditions, including character disorders (especially psychopathy), narcotics addiction, and alcoholism, may well have the relevant features as well. Though these areas provide extremely difficult cases, they too may involve the sort of 'defect of reason' or other general incapacity that precludes the requisite *mens rea* for criminal responsibility.

That a person suffers from psychosis of one sort or another, does not necessarily imply that on the occasion of acting they lacked either the knowledge or self-control that seem necessary to responsibility. Moreover, it is important to distinguish between knowledge and the *capacity* to reason. Even the most bizarre cases may still leave us wanting to say that the person knew what he was doing and could control his actions – in some sense at least. He may have taken precautions to avoid getting caught. He may have known what he was doing and even that it was wrong. Indeed, in the famous eighteenth-century case of Hadfield, it was a condition of his doing what he did that it be (legally) wrong:

Hadfield suffered from the delusion that he was destined to be another Christ, sacrificed by the state and a martyr to human salvation. In order to arrange his sacrificial death at the hands of the state, he purchased a gun and ammunition, arranged to be in a theater audience on an occasion when, as he knew, King George III would be attending; then, at a moment when the king was standing in his box in full view of the audience, Hadfield deliberately fired his gun in the direction of the king.[12]

Hadfield knew what he was doing, at least in a sense, and planned and plotted his action (that is, it was not done on mere impulse or under external compulsion) – nonetheless he was clearly suffering from "a defect of reason", clearly insane and not properly responsible for his actions. In saying he was 'psychotic', we must be understood as saying he was irrational: it is not just that his beliefs were false, he lacked the *capacity* to correct his understanding, to respond to his relevant reality. Again, psychiatrists may be able to help us understand the nature and extent of his psychological impairment, but there is more to understanding why that relieves him of criminal responsibility.

Which brings me back to my main point: mental disease, at least in relation to criminal insanity and responsibility, is not a purely medical notion. It is not a matter simply for experts to decide; it involves the sorts of questions of social and ethical value that ordinary lay people who serve on juries have equal competence to consider. How far are we willing to use people who could not help themselves for the purposes of the law? It is sometimes said that there would be no point in so using them, that punishing the insane would have no deterrent force because the insane are not rational calculators of the sort addressed by the law. However, even if it were true that it could not deter the individual offender or have reformative value for him, it is a mistake to think this shows it could have no value.[13] Punishing the insane might well deter many potential offenders – who might be especially impressed by the fact that *even* the insane get punished, that there is no way out. But the point is that such utilitarian value would be bought at the price of great individual injustice, a price greater than we are – in general – willing to pay.

VI

Where to draw the line, where it becomes unjust to pubish, depends on determination of criminal insanity, of "defect of reason from disease of the mind". And 'mental disease' is a boundary concept. It may help here to compare it with another cross-dimensional concept such as 'inadequate vision'. As Herbert Fingarette explains, how well a person sees is scientifically measurable. Oculists can establish in what ways and to what degree a person's vision deviates along a standard scale from some base point on the scale. But that does not make 'inadequate vision' into a purely medical concept.

> For example, Jones may have adequate vision to drive in a community that is willing to take great risks and that has poor public transportation; whereas he may have inadequate vision in a community where public transportation is excellent and cheap, and where public policy favors minimizing risk. In either case the medical condition of Jones is the same ([9], p. 39).

It is the task of the oculist to report on the condition, but it is the responsibility of the appropriate motor vehicle authority to set the standard. "The decision itself must derive from an authoritative interpretation of public policy in the light of technical information." Like 'inadequate vision', 'mental disease' can be understood as establishing a boundary within a dimension – the dimension itself may be the domain of medical expertise, in this case, psychiatric expertise, but the setting of the boundary raises legal and moral issues as well as issues of public policy, and these cannot be settled simply by reference to medical purposes. Insanity is a legal-moral boundary concept, and not a purely medical one.

What one wants from expert witnesses in an insanity defense trial is testimony about the nature and causes of any psychological incapacities from which an individual may suffer. Diagnostic labels and clinical conclusions are less important than the details on which they are based. One wants to bring into the court the broadest range of evidence possible about psychological state and capacity at the time of the crime. Which of course requires understanding of the defendant before and after the crime and at the time of the trial, at which time, because he has been found competent to stand trial, he may appear relatively sane. (In the insanity defense, there is usually no question but that the accused actually did the thing alleged, the question is whether the person who did it was the sort of person the law may properly hold responsible for his actions and punish.) Insanity is a legal standard rather than a medical fact and it is the place of the jury (in the end) to bring together medical information and social purposes.

VII

At this point it is worth emphasizing how difficult is at least one aspect of the task. Determining the powers and capacities of an individual is in itself exceedingly complicated and difficult – which is *not* to say impossible. That someone could have given a matter more thought, or with a different motive would have acted differently, may be as much a determinable fact about them as that a car has the capacity to do 100 mph.

> We may have as good ground for saying "well he just could not do it though he tried" as we have for saying "he didn't know that the pistol was loaded" ([14], p. 203).

But to see the difficulty and complications, consider for a moment the rather simpler case of the car. One may want to know of a given car at a given time whether it could do or could have done 100 mph. Suppose the car was just sitting still at the time. One could use one's knowledge of cars in general, the relation of horsepower to speed, and of engine construction to horsepower etc., and also one's knowledge of this car in particular, say, for example, its performance on past occasions – to answer the question. Suppose, however, that the car was being tested at the time. If the driver tried to press it to its limit and it nonetheless failed to do 100 mph, that might seem to settle the question of whether it could do 100 mph. But the question is in fact indeterminate unless one specifies which conditions are to be held constant and which are to be allowed to vary. Perhaps the car *could* do 100 mph *if* it were properly tuned. Perhaps if it had different tires. But how many changes would one allow before regarding it as a different car? That would depend on one's purposes, and especially on whether one's interest were in that particular car with its particular history and equipment on a particular occasion in particular circumstances (in which case a test drive might be decisive) rather than an interest in that car's total potential under a variety of conditions.

Determining the capacities of humans is even more complicated because of the role of desires – one may be able to run a mile in five minutes but simply not want to (if a car, by way of contrast, fails to do 100 mph in a road test, one cannot explain the failure by an absence of desire – unless the absence is in the driver). One must ask, inevitably, how much desires are in one's control. The truism that one can always try is no help. One cannot always alter or act against a given desire. On the other hand, not all of our desires are given, we can form deliberative desires, and we can subject given desires to criticism and – occasionally – change them. But again one must specify what is to be held constant and what allowed to vary. What is to be included in the controlling or essential 'self'? If everything, including external conditions, is held constant, it might seem that no one could ever have done anything other than what they in fact did on a given occasion. But we know better. We usually regard intentions as alterable, as subject to reason and responsive to relevant changes in facts or information. If you can give me a motive for acting differently, I can, in general, change my course of action. I often have both the opportunity and the ability to do things I do not in fact do. Suppose I went to a movie because I wanted to see the particular film which was playing. Had a different film been playing, I would not have gone. And no general thesis of determinism counts against this. The freedom that matters is not countercausal. I act freely when I act for certain sorts of reasons, not when I act for no reason or randomly. It would indeed be a picture of unfreedom if we were to discover that we are in fact governed by hidden inaccessible causes like persons under post-hypnotic suggestion. If a hypnotist orders a person under his power to open an umbrella upon wakening, but to forget the command, what happens is that the person opens the umbrella, and if asked why, makes up a reason. "To test it". If we rebut the reason: "But you have just used it, and it was all right." They will make up another reason, say "I am opening it for decorative effect". But their reasons are rationalizations. Their actions are independent of the reasons they produce. But we know, as well as can be known, that not all reasons as things are now are rationalizations. We are *not* all like hypnotists' subjects. We also know, or should know, that freedom (like responsibility) is a matter of degree; we are indeed sometimes moved by forces outside our knowledge and so outside our control. But we have nothing to fear from the advance of science – whatever we discover about the causes of human behavior we will not discover that reasons do not move. And that reasons move is all that we need in order to say we could have acted otherwise – at least sometimes. Our desires have causal efficacy. Where freedom is understood as self-determination and as a matter of degree (rather than, as certain existentialists would have it, all or nothing), we can see that knowledge can help us be more free; freedom is not a given, it is an achievement, sometimes a difficult achievement. Our personality and actions are shaped and influenced by many factors. This does not mean that as human knowledge advances and we uncover more and more of these factors we become less and less free and responsible. We may sometimes discover previous unfreedom, but that may be the necessary first step toward future freedom. There are many factors that operate in many different ways. In general, knowledge can help make one more free. Where before we were pushed about by forces of which we were unaware or that we did not understand, knowledge may put us in a position to intervene and so ensure that our actions match our intentions. For example, if we discover that too much alcohol always makes us violent or unsafe drivers, we can avoid alcohol and so violence, or avoid driving when drinking. (Indeed, rather than regard intoxication as a mitigating circumstance when someone kills through drunken driving, given what we know about the effects of alcohol, it might not be unreasonable to regard it as an aggravating circumstance, deserving perhaps a more severe penalty than mere reckless driving.) But, in general, the point is that given an understanding of the causes that influence us, we are in a better position to step back from those causes and so to control our own lives. We are then in a position to take steps, to interfere with the natural course of events – and if we decide not to interfere, that is itself a decision and we have thereby made the natural course of events our own and are therefore responsible (which may be part, at least, of what the existentialists mean when they say we are "condemned to be free"). Two further points at least should be noted before leaving this too brief and inadequate discussion of freedom. First, once one has stepped back, it is not always possible to intervene, we are sometimes up against a wall, and in those cases we are genuinely unfree, the victim of our conditions. Secondly, in order for knowledge to help make us free we must have it. That is, it is not enough for scientists to uncover causal connections – their knowledge must be disseminated and shared. Otherwise they and their employers, be they advertisers or government officials, are in a position to manipulate us, and the advance in knowledge (thus hoarded) would be movement towards an enslaving Brave New World rather than towards freedom.

I said that we know that not all reasons are rationalizations, that reasons sometimes genuinely move. But in determining in a particular case whether a person had the capacity to act rationally, we have to ask what is held constant and what regarded as alterable. What is included in the 'self' when one raises issues of 'self' determination? I have said that it cannot be absolutely everything. In the case of a person who commits a crime, the law sometimes asks "would he have done it had there been a policeman at his elbow". But this test perhaps places the stakes too high. Even kleptomaniacs and obvious psychotics might be able to abstain from action under the eyes of a policeman (their problems might not include a desire to be caught), and they still might not be able to control themselves, to alter their intentions in the relevant sense. The policeman-at-the-elbow test does not, in general, tell us what we want to know. Indeed, what we want to know may have more to do with the bizarre nature of the desires, their irrationality (their apparent incoherence and lack of interested motive) than with the person's ability to control them ([8], pp; 272–292; [21]). What we want to know is enormously complex and subject to many factors – too many for simple rules to be given. This is part of the reason that, in addition to expert testimony, the judgment of a jury is needed.

VIII

Even if what gets treated as a medical problem were not influenced by all sorts of extraneous factors, the existence of a disease (physical or mental) does not by itself excuse; some specific connection to the offense is needed, and this specific connection brings in the traditional, legally recognized grounds for excuse. Even while rejecting the 'product' formulation of the Durham rule, the D.C. Circuit Court in its 1972 Brawner decision insisted on such connection: "Exculpation is established not by mental disease alone but only if 'as a result' defendant lacks the substantial capacity required for responsibility. Presumably the mental disease of a kleptomaniac does not entail as a 'result' a lack of capacity to conform to the law prohibiting rape."[14] A person is not fully responsible for an action if he could not help it. As Aristotle pointed out, an action is less than fully voluntary if it is done in ignorance (i.e., because of mistake or delusion, etc.) or under compulsion. A general incapacity can go to show and to explain a particular failure of knowledge or control. The legal tests of insanity (M'Naghten emphasizing cognitive capacities and the ALI adding a control condition) refer to these conditions of voluntariness and so responsibility. 'Mental disease' serves as a general explanation of the lack of those particular capacities necessary for responsibility, and it is an explanation which suggests that the incapacities are not themselves culpable, are not self-created or 'chosen'. 'Mental disease' is not in and of itself an excuse. But a person who, due to mental disease, fails to distinguish a real gun from a toy gun, is no more to blame than a person who mistakes arsenic for sugar because of someone else's mislabelling. Ignorance changes the character of the subsequent actions and where the ignorance itself is not culpable, it excuses: for we then think the person could not help it, could not have done otherwise, had no alternative – under the circumstances was more like a cog in a machine than a free agent whose reasons inform and give character to his conduct. This depends, of course, on a particular view of the nature of mental disease: that it is something that *happens* to one, that its source is 'external' to the individual who suffers from it. Just as the ignorant mistake due to mislabelling is nonculpable only so long as the mislabelling is external and the acting on the label is not, in the circumstances, reckless or negligent, similarly, a person is not fully the author of his action if he acts under compulsion: if a person overwhelms you by force and makes you stab someone by pushing your arm, your arm and the knife become extensions of his arm, and the action is not yours but his. The situation is complicated because compulsion can operate through the mind instead of through physical force. If someone orders you to shoot or be shot, clearly you are acting under duress – whether this will count as an excuse will depend on who or what you end up shooting, the plausibility of the threat, etc. The situation becomes even more difficult when the compulsion is internal: when one is moved not by overwhelming external force but by overwhelming internal desire. How does one dissociate from an internal force? Not every succumbing to temptation is a surrender of the self, a passing of responsibility to an alien, irresistible impulse.[15] And it is not clear under what circumstances we should say that someone's capacity for control was diminished by mental disease, so that they could not help what they did, or that it was at any rate harder for them to help it. Freedom is a matter of degree. Much depends on which conditions are held constant (and by whom). How much is society willing to invest in changing conditions, how much of the self is regarded by the person as fixed? (Are we buying the car 'as is'? How much are we willing to put into new tires, etc.?)

In the light of our general framework, let us consider three concrete cases to clarify a few additional issues:

(A)

Defendant, an adult living at home with his parents, is a mentally deficient individual with an IQ in the imbecile range. He has been watching some neighbor boys play cops and robbers with capguns and upon finding a loaded revolver in the glove compartment of his father's car, he shoots and kills one of the neighbor boys. The patient's intellect is such that he does not discriminate between a capgun and a revolver, does not understand the meaning of death, does not know the difference between a television actor who falls down upon being shot and one of the neighbor boys who falls down pretending to be killed, and so forth. While in jail he asks whether the slain person will come to visit him. He shows no comprehension of what he has done or why he is involved with the police and is thoroughly baffled by the whole proceeding ([17], pp. 789 and 833–834).

Here clearly, the character of the behavior is different from what it would be for a person with ordinary capacities. While the harm is real, we would be misdescribing the action if we simply said the shooting constituted murder. Lightning may do real damage, but it does not 'comprehend' or 'intend' that damage. Our attitude towards the boy must be closer to that towards lightning than towards a human agent. He does not understand the nature and quality of his actions, and they therefore evoke a different response. He lacks the capacities needed to give his particular action the *mens rea* requisite to make it a punishable crime. His general disability provides a particular excuse for his behavior, but it also creates a general problem. While his behavior does not call for punishment, his condition calls for care and treatment, and others need to be protected from the special dangers his condition creates. Note that this would be so even if he had caused no harm: Whatever measures are called for, are called for by his condition, not his particular behavior (though his behavior is certainly good evidence of special dangers and the need for special measures). It does *not* follow that he has to be confined to an institution. Blackstone wrote:

In the case of absolute madmen, as they are not answerable for their actions, they should not be permitted the liberty of acting, unless under proper control; and, in particular, they ought not be suffered to go loose, to the terror of the King's subjects. It was the doctrine of our ancient law, that persons deprived of their reason might be confined till they recovered their senses, without waiting for the forms of a commission or other special authority from the crown . . . [3].

What measures are called for or justified must depend on the precise dangers feared, their likelihood in the circumstances, and the treatments or other precautions available.[16] In a case like this one, where the individual is not driven by bizarre or otherwise anti-social desires, but is merely mentally deficient and, in that way, like a child, perhaps no more is called for than the precautions one takes with a child. Certainly guns should not be accessible to either.

The main point to take hold of is that from the point of view of social policy, it is a mistake to consider questions of criminal confinement separately from issues of civil commitment and other forms of social control. The underlying problems and the institutional responses are tied. In the case of the insanity defense, at least since the English Act of 1800 "for the safe custody of insane persons charged with offences", the special verdict of "not guilty by reason of insanity" has been tied to continued confinement ([26], p. 78).

The insanity defense is statistically insignificant. Being excused on the insanity defense does not mean nothing happens; the consequences of a successful insanity defense may not be very different from the consequences of a criminal conviction. The stigma may in fact be greater; the total institution to which one is committed (the mental hospital) and the treatment one receives may be little better than in the typical total institution of punishment (the prison), and the sentence is indeterminate. As Abraham Goldstein puts it:

. . . insanity has become a defense in name alone. In virtually every state, a successful insanity defense does not bring freedom with it. Instead, it has become the occasion for either mandatory commitment to a mental hospital or for an exercise of discretion by the court regarding the advisability of such commitment. And because the commitment is for treatment, it continues until such time as the hospital authorities conclude the patient is ready for release ([12], p. 19)

which may be never. But a jury, lacking confidence in the curative and evaluative powers of psychiatry, may fear that it would be too soon, and so convict. This may have been one of the factors operating in the next case.

(B): Kemper, a huge man, killed his grandparents while still a teenager (fifteen) and had been committed to a mental institution. There he aided in psychological testing procedures and was himself eventually tested and released as sane. The record of his prior offense was sealed after he was twenty-one because he had been a juvenile at the time of commission. He proceeded to commit a series of bizarre and horrific rape-murders around Santa Cruz, California. He picked up a number of women hitch-hikers, killed them, raped them, then engaged in cannibalism and dismembered the bodies and dispersed the parts. He finally murdered his mother and a woman friend of hers and then turned himself in to the authorities. He expressed no remorse and exhibited no affect when recounting his actions. He was tried, found sane and judged guilty.

Here we have a person who understands the nature and quality of his acts – it is we who do not understand them. What are we to say of Kemper? His desires are bizarre, his actions horrific, and his reactions are inhuman; but he suffers from no delusions, is not psychotic and is not of limited intelligence; his main incapacity is one of control – the only clinical category that seems to fit him is 'psychopath.' Should he be considered legally insane?

The American Law Institute, in its insanity defense test, specifically provides a caveat meant to exclude 'psychopaths' or 'sociopaths': "the terms 'mental disease or defect' do not include an abnormality manifested only by repeated criminal or otherwise antisocial conduct".[17] There is great difficulty in gauging capacities for control. The fact that an impulse was not resisted (even if acted on repeatedly) does not show that it was irresistible. And the abnormality or bizarre nature of a desire is not evidence of its strength. As Lady Wootton puts it: ". . . it is by no means self-evident that the [wealthy kleptomaniac's] yearning for valueless porcelain figures is inevitably stronger, or more nearly irresistible, than the poor man's hunger for a square meal or for a packet of cigarettes".[18]

Bizarre desires also do not, in themselves, amount to mental illness. One should not hold that conformity is a condition of sanity. But some do want to hold that, and this is perhaps one of the things that leads those people to want to say that all criminals are 'sick'. Notice that engaging in prohibited behavior need not be a sign of an abnormal desire (especially not a desire to break the law as such, to do whatever it may happen to prohibit: to bite forbidden fruit precisely and only because it is forbidden). People commit all sorts of crimes for all sorts of reasons. That any crime may be shocking to some, does not make all crimes the product of abnormal desires nor all criminals sick. And even where one is dealing with a person who has clearly abnormal desires, one should clearly distinguish the aims and justifications of treatment and punishment. A compulsive exhibitionist, for example, is especially likely to violate indecent exposure statutes. When he does, he is liable for the criminal penalties attached to producing the sort of harm entailed. But that should not be a ground, as it is in certain 'sexual psychopath' statutes, for further indefinite confinement. Certainly it cannot justify confinement for purposes of treatment when no treatment is known or available. Indecent exposure statutes do not usually carry life sentences – the danger and harm are not great enough to warrant it. But that is in fact the effect of some sexual psychopath statutes. I would agree with those who would limit civil commitment under quasi-criminal statutes meant to deal with so-called 'sexual psychopaths' (the notion is in fact extremely vague and ambiguous) to the penal sentence for the offense actually committed.[19]

Returning to Kemper, we do not know where his bizarre and horrific desires come from nor how to treat him so that he would desire and be capable of resisting them. But in his case the penal sentence is for multiple murder; given that there is no known treatment, there is no point in confinement for treatment. Fear of premature release, especially in the light of his earlier release, might unsurprisingly weigh with the jury. In his case the damage and the danger are clear (the punishment is not for the mere oddness of his desires), and as Kemper himself is reported to have said: "If I were you, I would not let me loose."

That Kemper must be confined is clear. Whether he should be regarded as 'mentally ill' and therefore not responsible for legal purposes raises more difficult questions. Is he a psychopath? Is psychopathy simply a matter of repeated criminal actions? If it is more, ought it to diminish responsibility? To sort out these issues, we would have to go deeper into the questions about 'control' and the strength of desires already alluded to. It is clear in Kemper's case that there is more than just repeated crimes to consider: their character and his emotional response (or lack of it) to his acts must also be taken into account. Is he a person without a superego, and so incapable of experiencing guilt? Or is his incapacity perhaps less radical – perhaps more like that of habitual or hardened criminals (e.g., those raised among and associating only with criminals) who do not lack a superego and capacity for guilt altogether, though its objects are limited and perverse from society's perspective (so while believing in honor among thieves, it is only among thieves or to fellow thieves that there are felt to be questions of duty or honor)? Even if his incapacity were of the more radical sort, so he could not feel guilt about anything, what would that show about his responsibility and liability to punishment? What is the appropriate relationship of guilt or the capacity for guilt to punishment? Are penitentiaries only for the 'penitent' (or potentially penitent) – or may prisons appropriately serve other purposes? Why is it that there is a common death penalty statute which prohibits carrying out the sentence on an individual who develops mental illness *after* conviction? If a death penalty is ever appropriate, why does it become inappropriate once we think the person incapable of appreciating why it is being imposed? And if the feeling of guilt or the capacity for feeling guilt is a relevant consideration in punishment (and parole), how are we to deal with the evidentiary problems in a system that encourages (indeed, for purposes of appeal, typically requires) an individual to deny that he did the action and so to deny that he feels guilty about having done it?

In Kemper's case there is no alternative to confinement (and probably no alternative to confinement in a prison – assuming there is no special treatment a mental institution could offer and that he would be likely to raise special problems of security that prisons are in general better designed to deal with); but whether he should count as legally insane might still affect our attitude towards him (the difference between his being bad and being mad) and so to the sort of suffering he should be liable to. The mentally ill who are dangerous may require treatment and confinement, but treatment does not (necessarily) entail suffering. Punishment does entail condemnation and so (within our system and with our practices) suffering. The imposition of what kind and degree of suffering is justified by what? This takes one to the large question of the character of our penal practices, and specifically to the quality of prison conditions (e.g. should there be sexual deprivation in addition to deprivation of freedom of movement?), and what should be the relation of quality of prison life to quality of life in the wider society or among those who (through no fault of their own) are on welfare?

We cannot go into these questions here. But perhaps by way of summary we may note that all sorts of people do all sorts of things for all sorts of reasons – psychiatrists and other experts may help us in determining which reasons and causes actually moved in an individual case, but that does not settle what reasons should excuse or what steps society should or must take.

(C) And, finally, there is the case of Patty Hearst, who served a prison sentence for her participation in the Symbionese Liberation Army bank robbery. While not an insanity defense case, it raises many issues connected with the impact of psychological condition in the criminal law (role of motivation, role of expert witnesses, etc.), including the large issues of freedom and responsibility.

Her case is clearly different from that of other SLA members who participated in the bank robbery. While she asserted allegiance to their causes, they were her abductors, and she had been kidnapped. What difference does that fact make and why? Certainly there is no doubt that she participated in the robbery – the film evidence and the decision of the jury are conclusive on that point. But does she deserve punishment for her participation – does the history behind her involvement change its character?

The prosecution was concerned to show that Patty Hearst had been rebellious prior to and independent of her kidnapping. If she had been rebellious, one central issue could be understood in terms of the question whether her kidnapping 'intervened' or was merely a circumstance *through* which her rebelliousness fulfilled itself. On the latter view she exploited a fortuity, but her rebelliousness might have expressed itself in other ways anyway. On this view, her participation in the bank robbery could be presented as caused by her character, which was not in turn caused by her treatment. So she was responsible.

But where does character come from? Every member of the SLA had a history – there is a story behind the reasons that led them to join the SLA. Whether a particular story should exculpate someone is a complex and difficult matter. Being kidnapped is a dramatic event, sufficiently out of the ordinary to make it ludicrous to think of it as merely a background circumstance, even if one accepts the claim of an independent rebelliousness. My own view in the Hearst case is that she ought to have been forgiven something – precisely how much (whether it should have included her participation in the bank robbery, her subsequent shooting at the sporting goods store, etc.) is a properly disputable question. Certainly it could not provide an excuse for the rest of her life. The chain of consequences has to be cut somewhere (and if she had shot and killed someone during the bank robbery, it would be more difficult to allow her to disown that act). Again freedom is a matter of degree. There are certain sorts of personal history that rob people of the authorship of their actions – when one is coerced, one is not 'self-determining' or, in that sense, free. There are many types of coercion – it depends on what one includes in the original 'self' (do we start with what we think of as a rebellious Patty Hearst, or with a largely malleable and characterless Patty Hearst?), and how events operate on that self, and on what that self, finally, comes (or is led) to do. Patty Hearst's participation in the bank robbery looks very different if we think of it on a model of direct duress (she unwilling, very frightened, with gun pointed at her head), on a model of 'brainwashing' (she willing, but only because, in effect, a gun had been pointed at her head earlier transforming a self and instilling uncharacteristic beliefs), on a model of defensive identification with the aggressor (she willing, but only because internal guns have been called into defensive play). The jury rejected these models, and found her fully responsible. The point for us to see is the many dimensions of the self and the variety of factors that go into making an action more or less ours, more or less free.

Perhaps a word should be added about environment in relation to freedom. Confusion sometimes stems from the following two thoughts: (1) that environment and social deprivation are the main causes of certain sorts of crimes (e.g., people steal because they are hungry and unemployed); and (2) that this sort of external circumstance or environmental influence is incompatible with the 'free' choice of a life of crime. The first thing to notice is that the fact that our behavior (in general) has causes or sources is not (in itself) incompatible with its being ours or free. All of us have a history. A person's liking for music may have an explanation in the history of their upbringing and development – it is nonetheless a part of their character, and as they identify with it, it is a part of who they are. Certainly a liking for music does not receive social disapproval or punishment. What makes the actions I do out of my liking for music 'free' is not that the liking is uncaused. For a feature of my character to be fully mine it must have a certain sort of history, not no history (it may even include moments of external coercion – as when parents force their children to learn to play the piano). The difficulty comes in specifying the sorts of causes and reasons that are on the side of freedom.

Mental illness typically excuses. It is out of the person's control and typically causes them suffering and leads to a pattern of desires and life that is not satisfying. If abnormal behavior does not cause the agent harm, we tend to dismiss it as eccentricity, not insanity. And eccentricity is not an excuse, a person is considered responsible for it and its consequences. The problem arises when we ask what is so special about mental illness that makes it an excuse, and we say sweepingly: "the person is not free" or even more specifically, "cannot help what he does, he has no real alternative." For there are other ways to narrow alternatives: a person's environment and upbringing, their poverty and association with certain sorts of people, may limit their options. Why is it that we do not say under such conditions that people are not free and so not responsible (or less free and less responsible)? Perhaps they are not so responsible. Perhaps poverty should be an excuse for stealing – in a society and an economy that provide no alternative. One could imagine a defense on the grounds of 'economic necessity' ([14], p. 51). Whether or not our society is such a society and whatever the right attitude towards environmental limitation of alternatives as an excuse, notice that the recognition of environmental factors does not make all criminals 'sick' and crime the product of disease. I do not wish to go further into the character of 'disease' or the varieties of crimes and criminals and the reasons for committing crimes. What I want now to emphasize, is that confusion should not lead one to assume that a medical or treatment model would better deal with these problems where punishment may seem inappropriate. Treatment would still only be treatment of the individual, not of the source of the problem: the environmental source of his 'criminal tendencies'. What good does it do to alter his attitude to crime if society still leaves him and his fellows little alternative to crime? He may commit less crime, but punishment can also have that effect. 'Therapeutic treatment' does not automatically come to make sense with the recognition of the importance of the social sources of crime. There are two ways of changing beliefs about society: one is by operating directly on the beliefs, the other is by changing society. A man may want to believe that his son (whom he has not seen in years) is alive. On one view of his problem, it may be enough to hypnotize him; on another, it may be necessary to bring his son back (or even back to life) [29]. If society is concerned about crime and criminal attitudes or propensities, and we at the same time believe that society is itself the prime generator of crime and criminal attitudes and propensities, what we must do is change society. And for this, neither punishment nor treatment is likely to be the procedure of preference.

While insisting on objective standards in the criminal law, Holmes admits

> If punishment stood on the moral grounds which are proposed for it, the first thing to be considered would be those limitations in the capacity for choosing rightly which arise from abnormal instincts, want of education, lack of intelligence, and all the other defects which are most marked in the criminal classes ([15], p. 45).

The existence of the insanity defense may help legitimate incarceration of offenders who do not suffer from gross defects. It is worth emphasizing that so far as our special treatment of the mentally ill stems from their lack of certain capacities, we should recognize (as Holmes's list does) that incapacities can have other, external, sources. Environmental circumstances can limit opportunities and imaginable alternatives, and the person who steals from need may have as little choice as the mentally ill person who acts under delusion. The difference in the law's attitude may stem from the fact that we may seem to have an alternative treatment for the mentally ill (it may also be that people do not think of one's environment and upbringing as things that *happen* to one, that one suffers), but a recognition of the claims of the socially deprived would call not for a shift simply from punishment to individual therapy or treatment, but for change in wider social arrangements. The appropriate response would not be to condition the individual lawbreaker, but to change the conditions that led to his acts.

One final point: the insanity defense, I have said, is, for a variety of reasons, statistically insignificant. But understanding the rationale behind it is of enormous importance. For if we can understand why the insanity defense is a defense and to whom it applies, we may be in a better position to appreciate not only the present place of *mens rea* in the law, but also what is right and what is wrong with the suggestion that a medical or clinical approach should be taken to crime in general – that questions of *mens rea* and responsibility should be eliminated.

Many psychiatrists and social scientists suggest that all criminals should be regarded as sick, as suitable cases for treatment, but not punishment. Along with this, the argument is sometimes given that the purposes of the law would be more efficiently served if we treated crime, as we do disease, in its incipient stages – so that where we could predict that a person was likely to commit crimes (particularly violent crimes) we would start treatment or place them in preventive detention rather than wait for the damage to occur. I wish to suggest that there are grave dangers, as well as conceptual errors and confusions, in substituting too widely and quickly a 'treatment' model for a 'punishment' model in crime.[20] What may seem an enlightened and humane movement for reform, may in fact constitute an assault, a dangerous assault, on freedom and dignity. A concern for preserving the place of *mens rea* in the law may in fact be our best protection.

University of California at Santa Cruz

NOTES

1 Preparation of this paper was assisted by Faculty Research Funds granted by the University of California, Santa Cruz; and by support provided by the University of Texas at Austin, School of Law. I am also indebted to colleagues at Austin for helpful comments.

2 See Morris [20], especially pp. 105–109. Conspiracy laws and vicarious liability (as in the felony murder rule) raise other and difficult problems.

3 See Hart [14]. References to articles by Hart in subsequent notes will all be to their appearance in this collection.

Hart argues persuasively, throughout this collection, for a broadened understanding of the place and importance of *mens rea*. Emphasizing the operation of excusing conditions in the law, he shows they can be understood while rejecting the false dichotomy that would force one to choose between a preventive and retributive view of punishment. One can accept that beneficial consequences (prevention of harms and ensuring of conformity to law) are the general justifying aim of punishment, while nonetheless insisting that pursuit of that aim has to be limited by considerations of justice in the application of punishment to individuals. Those who could not have avoided breaking the law, where that incapacity is due to no fault of their own, are not suitable subjects for punishment.

More specifically, Hart points out that excusing conditions protect the individual and the value of individual choice against the claims of the rest of society (even if at some cost in the efficacy of our system of judicial threats). By recognizing excusing conditions (by demanding *mens rea*: knowledge of circumstances and foresight of consequences, or at least the capacity for foresight in cases of negligence):

First, we maximize the individual's power at any time to predict the likelihood that the sanctions of the criminal law will be applied to him. Secondly, we introduce the individual's choice as one of the operative factors determining whether or not these sanctions shall be applied to him. He can weigh the cost to him of obeying the law – and of sacrificing some satisfaction in order to obey – against obtaining that satisfaction at the cost of paying 'the penalty'. Thirdly, by adopting this system of attaching excusing conditions we provide that, if the sanctions of the criminal law are applied, the pains of punishment will for each individual represent the price of some satisfaction obtained from breach of law ([14], p. 47).

4 Chief Justice Brian of the Common Pleas in 1477, *Year Book*, 17 Pasch Ed. IV. f.1.p1.2. Quoted in Hart ([14], p. 188).

5 Wootton ([31], p. 74): "... since it is not possible to get inside another man's skin, no objective criterion which can distinguish between 'he did not' and 'he could not' is conceivable."

Wootton's skepticism is expressed in connection with the narrower issue of capacity for control, which we will return to. We can note here, however, that whatever the problems (of knowledge, evidence, and proof) she cannot avoid them. She wishes to confine consideration of a wrongdoer's state of mind at the time of the offense to a separate disposition phase of proceedings, but she cannot avoid altogether consideration of mental states and so the problems of proof she presses against consideration at an earlier phase. Even if one gives up punishment for treatment, the appropriate treatment for someone who causes harm by accident is surely different from that for someone who causes a similar harm on purpose.

6 See Kenny ([16], p. 148). In addition to considering whether panic is a 'reasonable' response to fear of apprehension, one must also consider whether it is relevant that the fear arises in the context of commission of a felony.

7 See Holmes ([15], pp. 48, 49, 50–51). It is worth reiterating Hart's point (note 3) that one can reject objective standards without rejecting a utilitarian view of the function of punishment: "Even if the general justification of punishment is the utilitarian aim of preventing harm, and not vengeance or retribution, it is still perfectly intelligible that we should defer to principles of justice or fairness to individuals, and not punish those who lack the capacity or fair opportunity to obey" ([14], p. 244).

[8] It is, of course, less objectionable where the liability is to a fine but not to punishment in the standard (condemnatory) form of imprisonment ([8], pp. 110–113). Strict liability in terms of fines or compensatory damages (at least) may make eminent sense in light of problems of proof, as an incentive to those engaged in socially desirable but hazardous activities to take the most precautions possible (and if they cannot, not to engage in the activity), and in terms of distribution of costs (even if the cost of total prevention is too high), it may be better that the agent creating the risk and profitting from it bear the cost of harms when they do – even unavoidably – occur [27].

[9] Recklessness involves a knowing disregard of unjustified risk. In negligence, one is typically not aware of the risk or danger created, but could have and should have been. *Mens rea* (sometimes defined as a matter only of knowledge of circumstances and foresight of consequences – which is all that is needed for oblique intention and so criminal liability) should be taken to include negligence, which while not itself a state or mind (the actor's mind may be 'blank') depends on certain psychological *capacities* and what the agent should have known, noticed, or done, given those capacities. See ([14], pp. 136–157).

[10] *Blocker v. U.S.*, 288 F.2d 853, 859–861 (D.C. Cir. 1961). Quoted in Stone ([24], p. 225).

[11] This is meant to be by way of contrast with physical disease, which is supposed not to involve such norms, but only the "structural and functional integrity of the human body . . . what [physical] health is can be stated in anatomical and physiological terms" ([25], p. 189). It is perhaps worth pointing out that the matter is more complicated than that. Certainly what counts as mental disease depends on norms (indeed, I am wishing to emphasize that claim especially in relation to *legal* insanity), but the definition of physical disease too can depend on social and ethical norms:

> there is in fact very little disagreement among us over what constitutes the proper working of the human *body*. We all would agree that a body with paralyzed limbs was no more in 'good working order' than a car with flat tires; and, in general, our culture identifies bodily health with vigor and vitality. But we can imagine a society of mystics or ascetics who find vitality a kind of nervous distraction (much as we regard hyperthyroid activity) – a frustrating barrier to contemplation and mystic experience and a source of material needs that make constant and unreasonable demands for gratification. Such a group might regard bodily vitality as a sickness and certain kinds of vapidity and feebleness as exemplary health. Our disagreement with these people clearly would not be a purely medical matter ([8], pp. 254–255).

See also J. Glover [11] and Anthony Flew [10], especially pp. 26–94.

[12] See ([9], p. 138; also [26], Chapter 4). That Hadfield was suffering from a war wound to his head which led to fits of mad behavior and that left the membrane of his brain exposed for the jury's inspection must have doubtless also been impressive.

[13] It is also not the case that the mentally ill as a class are not responsive to fear of punishment. On the contrary, discipline in mental hospitals typically depends on just such fear. See *Durham v. U.S.* 214 F.2d 862 (D.C. Cir. 1954), n. 30; see also ([14], pp. 18ff, pp. 40ff) and ([13], p. 471): "What distinguishes the sane from the insane under criminal law standards is the inability to appreciate the *culpability*, not the punishability, of their conduct."

[14] *U.S. v. Brawner*, 471 F.2d 969, 991 (D.C. Cir. 1972), Cf. [8], p. 273; [28], pp. 330ff.

Herbert Packer, on the contrary, seems to claim that no particular specific connection need be made, that a general condition is all that is required ([23], p. 135):

The idea there [in the peculiar defense of 'partial insanity' or 'diminished responsibility' in homicide] is that mental illness might be adduced to show that the actor lacked the state of mind required for a particular degree of homicidal crime as, for example, premeditation and deliberation in first-degree murder. That would have the effect, if accepted by the trier of fact in the peculiar case, of reducing what would otherwise be first-degree murder to second-degree murder or even manslaughter. This defense bears no relation to the insanity defense, which is not at all addressed to particular elements of the offense but rather to the actor's general mental condition. The 'partial insanity' defense is simply a useful ploy to avoid conviction for an offense that might incur the death penalty.

In sum, the insanity defense is not implied in or intrinsic to the complex of mental element defenses that make up most of the law of culpability. It is an overriding, *sui generis* defense that is concerned not with what the actor did or believed but with what kind of person he is.

[15] Consider J. L. Austin's famous footnote [2]:

I am very partial to ice cream, and a bombe is served divided into segments corresponding one to one with the persons at High Table: I am tempted to help myself to two segments and do so, thus succumbing to temptation and even conceivably (but why necessarily?) going against my principles. But do I lose control of myself? Do I raven, do I snatch the morsels from the dish and wolf them down, impervious to the consternation of my colleagues? Not a bit of it. We often succumb to temptation with calm and even with finesse.

[16] There are large and difficult issues connected with preventive confinement, 'dangerousness', and prediction. Alan Dershowitz provides a useful perspective on many of these issues in a number of his writings; e.g. [4, 5, 6, and 7]. See also Stone [24].

[17] See [1]. See also *U.S. v. Brawner*, 471 F.2d 969 (D.C. Cir. 1972) at 992–994; *Wade v. U.S.*, 426 F.2d 64 (9th Cir. 1970); and *U.S. v. Currens*, 290 F.2d 751 (3rd Cir. 1961) – all of which raise questions regarding the ALI provision concerning psychopathy.

[18] See ([30], p. 235. In her book review [32], Wootton goes further:

... it is clear that in certain circumstances the limits of a man's knowledge and understanding can be convincingly demonstrated. Thus, if I am asked to translate a passage from Japanese into English it is indisputable that this is beyond my powers: everyone knows that merely trying harder will not make me any more successful. But if I assert that I have an uncontrollable impulse to break shop windows, in the nature of the case no proof of uncontrollability can be adduced. All that is known is that the impulse was not in fact controlled; and it is perfectly legitimate to hold the opinion that, had I tried a little harder, I might have conquered it. It is indeed apparent that some people, such as sadistic sexual perverts, suffer from temptations from which others are immune.

But the fact that an impulse is unusual is no proof that it is irresistible. In short, it is not only difficult to devise a test of volitional competence the validity of which can be objectively established: it is impossible.

But there are other sorts of evidence (past behavior, action when other incentives were given – hence 'policeman at elbow' test – etc.). And similar problems arise in testing cognitive *capacity*: IQ tests are notoriously problematic, and all tests of human potential (as opposed to achievement) are complicated by the place of desire. And, finally, whatever the problems, Wootton cannot avoid them by her proposals because, as we have noted (note 5), they would have to come up again at the disposition phase of proceedings.

J. Feinberg provides a very useful discussion of bizarre desires and problems of strength and control in ([8], pp. 282–292). One should also recall in this context Austin's point about succumbing to temptation not necessarily being the same as losing control (note 15).

[19] E.g., Stone ([24], Chapter 11). See also *Cross v. Harris,* 418 F.2d 1095 (D.C. Cir. 1969).

[20] For some of the relevant arguments, see Hart ([14], pp. 186–209), and Morris [19].

BIBLIOGRAPHY

1. American Law Institute, *Model Penal Code*, Proposed Official Draft, sec. 4.01 (1962). [First presented in 1955].
2. Austin, J. L.: 1956–57, 'A Plea for Excuses', *Proceedings of the Aristotelian Society* 57, 24 (fn. 13).
3. Blackstone, W.: 1897, *Commentaries on the Laws of England*, Book IV, Rees Welsh, Philadelphia, p. 1442, *25.
4. Dershowitz, A.: 1968, 'Psychiatry in the Legal Process: A Knife that Cuts Both Ways', *Trial* 4, 29–33.
5. Dershowitz, A.: 1970, 'The Law of Dangerousness: Some Fictions about Predictions', *Journal of Legal Education* 23, 24–56.
6. Dershowitz, A.: 1973, 'Preventive Confinement: A Suggested Framework for Constitutional Analysis', *Texas Law Review* 51, 1277–1324.
7. Dershowitz, A.: 1974, 'The Origins of Preventive Confinement in Anglo-American Law', *Cincinnati Law Review* 43, 1–60 and 781–846.
8. Feinberg, J.: 1970, *Doing and Deserving*, Princeton University Press, Princeton.
9. Fingarette, H.: 1972, *The Meaning of Criminal Insanity*, University of California Press, Berkeley.
10. Flew, A.: 1973, *Crime or Disease?*, Macmillan, New York.
11. Glover, J.: 1970, *Responsibility*, Routledge and Kegan Paul, London.
12. Goldstein, A. S.: 1967, *The Insanity Defense*, Yale University Press, New Haven.
13. Gross, H.: 1975, 'Mental Abnormality as a Criminal Excuse', in J. Feinberg and H. Gross (eds.), *Philosophy of Law*, Dickenson Publ. Co., Encino, California, pp. 466–476.
14. Hart, H. L. A.: 1968, *Punishment and Responsibility*, Oxford University Press, Fair Lawn, New Jersey.

15. Holmes, O. W., Jr.: 1881, *The Common Law*, Little, Brown and Company, Boston.
16. Kenny, A.: 1968, 'Intention and Purpose in Law', in R. S. Summers (ed.), *Essays in Legal Philosophy*, Blackwell, Oxford, pp. 146–163.
17. Livermore, J. M. and Meehl, P. E.: 1967, 'The Virtues of M'Naghten', *Minnesota Law Review* **51**, 789–856.
18. MacIntyre, A.: 1971, 'Emotion, Behavior and Belief', in his *Against the Self-Images of the Age*, Schocken, New York, pp. 230–243.
19. Morris, H.: 1968, 'Persons and Punishment', *The Monist* **52**, 475–501.
20. Morris, H.: 1968, 'Punishment for Thoughts', in R. S. Summers (ed.), *Essays in Legal Philosophy*, University of California Press, Berkeley, Los Angeles, pp. 95–120.
21. Morris, H.: 1974, 'Criminal Insanity', (review discussion of Fingarette), *Inquiry* **17**, 345–355.
22. Neu, J.: 1975, 'Thought, Theory, and Therapy', *Psychoanalysis and Contemporary Science* **4**, 103–143.
23. Packer, H.: 1968, *The Limits of the Criminal Sanction*, Stanford University Press, Stanford, California.
24. Stone, A.: 1975, *Mental Health and Law: A System in Transition*, National Institute of Mental Health, Rockville, Maryland.
25. Szasz, T.: 1973, 'The Myth of Mental Illness', in J. G. Murphy (ed.), *Punishment and Rehabilitation*, Wadsworth, Belmont, California, pp. 186–197.
26. Walker, N,: 1968, *Crime and Insanity in England*, Vol. I, University of Edinburgh Press, Edinburgh.
27. Wasserstrom, R.: 1960, 'Strict Liability in the Criminal Law', *Stanford Law Review* **12**, 730–745.
28. Wasserstrom, R.: 1972, 'Why Punish the Guilty?', in G. Ezorsky (ed.), *Philosophical Perspectives on Punishment*, State University of New York Press, Albany, New York, pp. 328–342.
29. Williams, B.: 1973, 'Deciding to Believe', in *Problems of the Self*, Cambridge University Press, New York.
30. Wootton, B.: 1959, *Social Science and Social Pathology*, Macmillan, New York.
31. Wootton, B,: 1963, *Crime and the Criminal Law*, Stevens and Sons, London.
32. Wootton, B.: 1968, 'Review of *The Insanity Defense* by Abraham S. Goldstein', *Yale Law Journal* 77, 1019 and 1026–1027.
33. Zimmerman, P. D. and Goldblatt, B. (eds.): 1968, *The Marx Brothers at the Movies*, Putnam Berkley, New York, p. 112.

JULES L. COLEMAN

MENTAL ABNORMALITY, PERSONAL RESPONSIBILITY, AND TORT LIABILITY

This essay will explore the applicability of tort liability to the mentally defective or abnormal. The common law holds that the mentally defective are at fault for their torts. Does this rule, however, square with our ordinary conceptions of justice and personal responsibility? My view is that it does, for the most part, and, moreover, that it ought to, since in general the common law rule makes good sense. I will argue that there is a perfectly defensible analysis of the concept of fault according to which many categories of the mentally impaired are justly held liable for their faulty behavior. While I will argue that mental abnormality in general ought not defeat an ascription of fault in torts, I will also contend that considerations of mental deficiency ought not to be abandoned entirely in determining liability. My view is that mental deficiency may surface as a defense in torts not to deny fault, but to deny that the minimal conditions of agency or action have been satisfied. First I want to argue that neither the traditional nor the prevailing accounts of fault can explain the common law rule.

I. THE PROBLEM

There are two basic questions in tort theory: (1) On what grounds is a victim entitled to recompense for his losses? (2) When is an injurer obligated to render compensation to his victims? That these two issues are conceptually distinguishable is illustrated by the fact that one can imagine a social insurance scheme of tort law according to which victims would be compensated not by their injurers but through general tax coffers. Accordingly, accident *victims* could be compensated, though no *injurer* would be required to pay any more into the coffer than would any other individual in the same tax bracket. Whether such a system comports with our ideals of justice and efficiency remains to be worked out. The point is simply that those considerations which ground an entitlement to recompense need not coincide with those that ground an obligation to compensate.

However plausible it may be to distinguish the grounds of recovery from those of liability, most legal theory and practice does not.[1] As a general rule, liability and recovery are decided on the basis of the fault principle. That is,

B. A. Brody and H. Tristram Engelhardt, Jr. (eds.), Mental Illness: Law and Public Policy. 107–133.

a victim is entitled to recompense only for those losses that are the fault of another. In the absence of misconduct on the part of an injurer, the victim is not entitled to recompense, though it may be sound public policy to compensate him. Compensating him may be further warranted by considerations of welfare, utility or benevolence. Nevertheless, compensating the victims of faultless conduct – in most cases – does not appear to be required by those particular principles of justice that govern the law of torts.

Liability for the most part is also governed by the principle of fault. An injurer is obligated to render compensation only for those harms that are his fault. The principle of fault liability, however, has been largely abandoned in the liability of manufacturers for their products, nuisance law, workman's compensation (which is statutory, not common law), ultrahazardous activities, and, to a lesser extent, in (no-fault) automobile accident law. Because victims and injurers usually have their respective claims adjudicated within the fault framework, these departures from the general rule are naturally, but wrongly, thought to involve a necessary if somewhat undesirable trade-off of justice for utility [3].

Continuing efforts to abandon fault as a basis of liability and recovery and subsequent doubts about the justifiability of doing so have drawn attention to the concept of fault itself. For while it has been the common legal practice to decide liability on the basis of fault, it has been considerably less common for legal theory to analyze the standard or to justify its use. Two major theories about the meaning of fault and the justification for imposing liability on the basis of fault have surfaced. They are what I will call The Theory of Responsibility and The Theory of Loss Allocation.[2] Both theories share the view that liability (and recovery) ought for the most part to be decided on the basis of fault; or, in other words, that the absence of fault should be a defense to liability. That is where the similarity ends. According to the former, liability is imposed to punish or thwart moral deficiency; according to the latter, it is imposed to minimize economic inefficiency. The question is whether the mentally abnormal are suitably immoral or inefficient.

II. THE THEORY OF MORAL RESPONSIBILITY

It is a commonplace that in torts liability is based on the principle of fault. It has not always been that way. As every first year law student knows – indeed every first year torts course seems to be woven around this historical tale – liability was originally determined without regard to fault.[3]

During the 13th and 14th Centuries the number of *writs* available in the

King's Common Law Courts was limited. If a victim could not fit his cause of action into the rigid form of an already existing writ, he was without grounds for litigation. At first the only available tort writ was Trespass (*vi et armis*), which established a strict liability case for recovery. An action for Trespass would succeed if the victim could establish that the harm of which he complains is a direct causal consequence of the defendant's conduct. (1) Defendant conduct, (2) victim harm, and (3) a direct causal relation between the two constituted the formula for liability and recovery under Trespass; it was all that could justify an action for damages. Evidence of the defendant's fault was unnecessary to establish a case for recompense under Trespass; accordingly, absence of fault was apparently insufficient to defend against a judgment of liability.

It was under the old writ of Trespass that the first important case in which mental illness was discussed as a defense arose; the case was *Weaver v Ward*, 80 E.R. 284 (1616). It appears from the report of the case that in the course of a military exercise the defendant discharged his musket, inadvertently wounding the plaintiff. The court, applying the Trespass formula, found for the plaintiff, holding that inadvertence was no defense under Trespass.

The *Weaver* court then went on to discuss mental defect as a defense, and though their discussion was only *dicta,* it formed the basis of a line of cases reaching similar conclusions:

> . . . for though it were agreed that if men tilt or turney in the presence of the King, or if two masters of defence playing their prizes kill one another, that this shall be no *felony*; or if a lunatick kill a man, or the like, because felony must be done *animo felonico*: yet in *trespass*, which tends only to give damages according to hurt or loss, it is not so; and therefore if a lunatick hurt a man, he shall be answerable in trespass . . . (*Weaver v. Ward,* at 284).

Arising as it did within the context of the *strict liability* writ of Trespass, the *Weaver* decision was a somewhat unlikely basis for subsequent decisions against the defense of mental illness under the *fault* doctrine. According to *Weaver* and the Trespass doctrine generally, mental illness or defect was not really distinguishable from other excusing conditions one might offer, for example, duress or ignorance. The writ of Trespass apparently did not admit excusing conditions and mental abnormality was simply no exception to that rule.

In general, Trespass constituted a very artificial basis of recovery. Victims of 'indirect' injury could not.[4] Eventually individuals who were injured indirectly could secure a cause of action under an additional writ of Trespass on the Case, or, more simply, Case. Because Case was an action for *all* indirect harms, some additional requirement had to be met to distinguish indirect harms which were compensable from those which were not. Failing that, all harms would have been subject to remedy, whether directly or indirectly caused, and the distinction between Trespass and Case would have been useless. The new element in the Case formula was fault, or at least what we have now come to regard as its earliest ancestor. To seek and expect to receive compensation under Case, a victim would have to show that the defendant's action was in some suitably narrow sense negligent or at fault.

Rather than obviating the inequities of the Trespass writ, the writ of Case compounded them. Consider an example. Suppose we have an injurer, *X*, and two of his victims *Y* and *Z*. *Y* is a victim of a direct harm, so he brings action under Trespass; *Z* is victimized by an indirect harm, so he seeks the writ of Case. If *X* is not at fault in the appropriate sense, *Z* cannot recover, but *Y* can! That is, *X* would be liable to *Y* but not to *Z*. Having established that he undertook to act with reasonable foresight and due care, ought not *X* be relieved of liability to *Y* as well as to *Z*? Ought not his reasonable care or the accidental nature of the occurence outweigh whatever measure is to be given to the directness of the hit?

These questions were not adequately addressed within context of Trespass and Case, yet their poignance could neither be denied nor merely overlooked. Eventually, the old framework caved in under the weight of these sorts of difficulties. But it was not until the landmark case of *Brown v Kendall*, 6 Cush. 292 (1850) that the principle of fault took hold as the standard of liability in cases both of *direct* and *indirect* harm.

In *Brown*, the defendant, in an effort to separate two fighting dogs, one belonging each to him and the plaintiff, beat them with a long stick. The plaintiff approached the dogs, both of whom began moving towards him. The defendant with his back to the plaintiff, retreated continuing to beat the dogs. While retreating, he unwittingly, but *directly* hit the plaintiff in the eye severely injuring him. Nevertheless, the plaintiff was denied recovery, since he was unable to show that in injuring him, the defendant acted in a negligent manner.

Brown v Kendall suggests that the directness of the hit lacks moral significance. Indeed, considerations of the directness of the hit had come to represent little more than a stumbling block to equitable adjudication of claims. With *Brown*, the standard of Trespass with its roots in strict liability finally gave way to a *moral* foundation of both recovery and liability. The simple causal question of Trepass was argumented by complex questions raised by the fault requirement; concern shifted from the order of events to the character of the defendant's conduct.

On the orthodox view, therefore, *Brown* marks a dramatic shift from the non-moral standard of strict liability to a moral theory of tort liability according to which the plaintiff's right to recover is determined by his ability to establish the *moral responsibility* of his injurer. On this view, to be at fault is thought to represent a moral shortcoming or defect in one's conduct. Legal fault is grounded in moral fault in the sense that those at fault in torts are morally responsible for the damages they cause. Tort liability is justifiably imposed either to punish (or penalize) or thwart that moral deficiency or to compensate for the losses of its victims. In addition, a judgment of fault is thought to stain the records of those held liable and is imposed not only to identify and punish moral fault but to express appropriate moral sentiments towards the faulty.

Certainly that is the account both of the development and meaning of fault as a standard of tort liability most of us have come to embrace as gospel. Unfortunately, it is a seriously mistaken one [4].

Brown v Kendall in the United States and *Holmes v Mather* in England introduced fault as the primary standard of tort liability, yet the concept of fault they introduced is very different from our moral one. For it is apparently a necessary condition of moral fault that the actor fail to comply with a standard of conduct within his reach; that he fail where he could have succeeded. From the onset, the standard of fault in torts was considerably more relaxed, as the case of *Vaughan v Menlove*, 3 Bing. N.C., 468 illustrates. In that case, the court held that a genuine (*bona fide*) effort to do one's best is no defense should one's conduct fall below the standard of care attributable to a reasonable man. It is perfectly plain then that one may be at fault in torts in doing the best one can. And if we feel that at least some mentally abnormal persons are genuinely incapable of measuring up to the higher standard of care, they may be both legally at fault and morally blameless. I do not mean to suggest that mental illness always indicates a lack of moral culpability. I am not one to romanticize mental illness that much. What I do mean to be saying is that conceiving of fault in torts as grounded in notions of moral responsibility does not adequately explain the liability of the mentally deficient who cause harm through no fault of their own, that is, who despite genuine efforts to do the best they can fail to measure up to our standards of care.

Richard Posner makes the same point more generally:

Characterization of the negligence standard as moral or moralistic does not advance analysis. The morality of the fault system is very different from that of everyday life. Negligence is an objective standard. A man may be adjudged negligent though he did his best to avoid an accident and just happens to be clumsier than average. In addition, a number of rules of negligence liability are hard to square with a moral approach. Insane people are liable for negligent conduct though incapable of behaving carefully ([15], p. 31).

And Holmes makes the same point rather conclusively:

If, for instance, a man is borne hasty and awkward, always having accidents and hurting himself or his neighbors, no doubt his congenital defects will be allowed for in the courts of Heaven, but his slips are no less troublesome to his neighbors than if they sprang from guilt neglect. His neighbors accordingly require him, at his proper peril, to come up to the standard, and the courts which they establish decline to take his personal equation into account [11].

Thus, it would be very hard indeed to square the common law liability of, for example, the severely retarded with the view that tort liability is grounded in principles of moral responsibility.

To the question, "What's special about mental illness?" the answer in torts is apparently 'Nothing'.[5] If anything, mental impairments are treated with greater skepticism and less respect than are physical shortcomings. For a blind man is expected to exercise the care of a reasonable *blind* person; whereas, an emotionally disturbed person is required to comply with the standard of a reasonable and emotionally stable person.

Mental and other shortcomings may prevent an actor from complying with the standard of due care, perhaps because the defect is a cognitive shortcoming which interferes with his capacity to appraise risks or to choose courses of conduct in the light of their relative riskiness. Or it may be that the actor's physical disability makes it impossible for him to carry out his intentions skillfully and carefully. In any case, defects of mental and physical skill, of emotion and comprehension may defeat an allegation of moral blame by providing evidence that the actor's noncompliance is not the result of defective character or motivation. Lack of moral fault is no defense, however, to tort liability. Fault, especially negligence, is conduct, not a state of mind. Scrutiny into the inner facts, weighing the 'personal equation', essential in fixing moral blame, is irrelevant in torts.

That lack of moral fault is no defense to civil liability proves that the concepts of legal and moral fault fail to coincide. If the concept of fault in torts is not a moral one in this sense, what kind of concept is it? And if legal fault is a more relaxed standard of conduct than is the standard of moral fault, what moral argument can one provide to justify employing it as a basis of liability?

III. THE ECONOMIC ANALYSIS

Having abandoned the view that the concept of fault measures the moral culpability of injurers (or victims), recent work in torts has taken two distinct paths. One group of theorists has attempted to breath new life into the view that torts is grounded in moral principle; others have taken their cue from economic theory. I want to consider the economic theory of torts first.

The economic analysis of torts is grounded in neo-classical welfare economics. To understand the relationship between law and economics, it is important to keep in mind that one traditional purpose of legal rules is to structure and protect an allocation of resources. Legal rules do so by conferring initial entitlements to resources, by restricting or defining the uses to which they can be put, by providing ground rules for transferring them, and by imposing liability, either criminal or civil, for unwarranted interference with them. Economic theory is also concerned with an allocation of resources and the uses to which they are put. According to classical welfare economics, the standard for judging an allocation of resources is its efficiency.

A distribution of resources is maximally efficient when it is Pareto Optimal. Resources are arranged in a Pareto Optimal fashion when any further rearrangement of them will make someone better off *only* by making at least one person worse off. So a Pareto Optimal or efficient distribution exists when no one can be made better off without making someone worse off – as judged by each person's standard of welfare or satisfaction.

Value is maximized and efficiency promoted most often (in theory) in markets. Resources are put to use by those persons who want them most – those who are made happiest by having them. One's desire for a resource is, in the market, measured by one's willingness to pay for it. Were it not for market failures, that is, lapses in the market's ability to promote efficiency or maximize value, legal rules would be, from the economic standpoint, redundant. Legal rules are justified when (and perhaps only when) they overcome market failures.[6]

The first systematic attempt to analyze legal rules in classical economic terms is Ronald Coase's masterful piece: "The Problem of Social Cost" [2]. Coase's point is a simple one and his argument for it elegant. Were we to assume both the perfect rationality, and knowledge of the participants in the process as well as the absence of transaction costs, any initial allocation of entitlements to a resource will yield a Pareto Optimal result. Put another way, the choice of initial entitlements would be neutral to the goal of efficiency.

Suppose a railroad seeks full use of its right of way. Full usage implies freedom to emit sparks without liability. Now suppose a farmer owning adjacent land seeks full use of his land for the purposes of growing crops. The choice is to entitle either the railroad or the farmer to the use of the land directly adjacent to the tracks. Suppose further that the value of that usage is $ 100.00 to the railroad and $ 50.00 to the farmer. If the farmer is entitled to the usage, the railroad will buy it from him at a negotiated price between $ 50.00 and $ 100.00. Both parties will be made better off by the sale and the initial allocation with the zero transaction cost assumption will lead to a Pareto Optimal result. On the other hand, if the railroad is initially entitled to the land use, no transaction will occur, because the railroad will not sell for less than $ 100.00 and the farmer will not offer more than $ 50.00. Again the result is Pareto Optimal, though a different Pareto Optimal point is reached with quite different wealth distributional consequences.

The Coasian world bears little resemblance to the one in which legal rules shape and fix behavior. Coase's theorem is then only a guideline to legal rule-making. The major point at which the Coasian world and our world diverge is with respect to transaction costs. Transactions are often costly and, on occasion, nearly impossible to make. Because transaction costs threaten efficiency, legal rules should be designed to limit them. For the most part, that is done by conferring initial entitlements to those individuals who, had they *not* been initially entitled, would have bought the 'right' away from the party upon whom it was initially conferred. As a corollary, liability should be imposed on those individuals who, were they initially entitled, would have *sold* it away to the highest bidder. Rights are therefore conferred on efficient uses and liability on inefficient uses.

When translated to torts, the economic analysis maintains that an individual is at fault when the costs of his activity outweigh its benefits. To be at fault is to engage in conduct the social costs of which outweigh the personal costs of accident *prevention* or *curtailment*. It is, in brief, to act in an economically inefficient manner.

To comply with the reasonable man's norm of conduct is to engage in accident prevention when the cost of doing so is less than the social costs of the accident should it occur; it is also to forbear from taking such safety measures when doing so involves a greater cost than the social costs of the accident. To be at fault in causing harm is to cause an accident that is not worth its social costs.

Proponents of the economic analysis urge that this is in fact the test of fault applied in negligence cases. As evidence for this contention, they cite the now famous Learned Hand formulation of the negligence calculus:

. . . a defendant is guilty of negligence if the loss caused by the accident, multiplied by the probability of the accidents occurring, exceeds the burden of precautions that the defendant might have taken to avert it. (*United States v. Carroll Towing Co.*, 159 F2d, 169 (1947)).

In explaining the Hand formula, Posner writes:

Hand was adumbrating, perhaps unwittingly, an economic meaning of negligence (fault). Discounting (multiplying) the cost of an accident if it occurs by the probability of its occurrence yields a measure of the economic benefit to be anticipated from incurring the costs necessary to prevent the accident. If the cost of safety or of curtailment – whichever cost is lower – exceeds the benefit in accident avoidance to be gained by incurring that cost, society would be better off in economic terms, to forego the accident prevention . . . Furthermore, overall economic value or welfare would be diminished rather than increased by incurring a higher accident-prevention cost in order to avoid a lower accident cost ([15], p. 32).

The orthodox view apparently fails to explain the liability of the mentally deficient because its view of fault as derived from our notions of moral culpability or responsibility does not square with the plain fact that the mentally abnormal are often morally blameless for their conduct. The economic analysis obviates this problem. For while the extremely mentally deficient may lack moral responsibility, it does not follow that their conduct cannot be economically inefficient. So while the ill or the incapable may fail to be morally culpable actors, they may be very inefficient ones. And they are held liable in torts for their inefficiency, not for their moral shortcomings.

The economic analysis *may* explain what we mean by fault, yet it does not follow that it provides an adequate justification for deciding claims of liability and recovery on the basis of fault.

Liability on the economic theory is imposed as an incentive to minimize faulty, that is, inefficient conduct. The analysis is therefore entirely forward-looking. The costs of any particular accident are 'dead'. Determining where, if anywhere, the fault lies has value only insofar as it enables us to locate upon which litigant the economic incentive either to take accident-preventive measures or to curtail an activity ought to be imposed. Moreover, identifying the party upon whom the economic incentive to take precautions ought to be imposed is important only to the extent that the party who is the cheapest cost-avoider is a member *of a general class of persons* engaging in or likely to engage in similar inefficient behavior. Holding him liable is a signal to them to take precautionary measures or to curtail their activities entirely. His liability is supposed to provide *them* with a *good reason* for avoiding undesirable harmful behavior. It is in this sense that the economic analysis might be thought of as a pricing theory of *general deterrence.*

The economic analysis provides one perfectly plausible (at least defensible) sense in which everyone, including the mentally and physically handicapped, may be at fault in acting. One is at fault whenever one undertakes to perform an inefficient act: that is, one that is socially more costly than beneficial. It does not follow from this instrumentalist conception of individual fault that imposing liability on the basis of fault will promote the goal of efficiency. Though we might agree that even the mentally deficient are (generally) able to form intentions and act on the basis of reasons, it hardly follows that their actions will be shaped by economic incentives.

The point can be made more generally. The economic analysis creates a pricing model for general deterrence. Individuals discover the prices fixed to alternative courses of action which reflect their accident costs, and decide which act to perform by weighing relative costs and benefits. The question is whether the pricing system is an effective tool in shaping the behavior of the mentally impaired.

Before the pricing model can work to minimize accident costs certain minimal conditions must be satisfied. The category of persons who are to be deterred by the liability decision must: (1) be able to *identify* themselves as a member of the appropriate category; (2) understand that they too are subject to liability for engaging in conduct similar to the defendant's, and (3) consider the likelihood and severity of liability a *good reason* for acting in compliance with the relevant norm.

Where mental defect affects adversely one's cognitive skills and thereby restricts one's ability to recognize similarities or to generalize from particulars, the economic justification for imposing liability is questionable. Further, where one is unable to understand that likely sanctions are supposed to be grounds for modifying one's conduct, the economic argument fails to justify imposing liability. Failure to appreciate causal relations, laws of nature, and an inability to recognize family resemblances are among the distinguishing characteristics of certain levels of mental retardation. In such cases, the action may be at fault, i.e. inefficient, though there is no reason to believe that imposing liability will promote general deterrence.

One need not suffer from a cognitive impairment for the economic argument to fail to present a good reason for imposing liability. Certain mentally *ill* persons are also unlikely to be moved by economic incentives.[7] The extremely compulsive personality may pursue a course of action in spite of its harmful consequences to others: regardless of the sanctions that may be imposed upon him. Sociopathological individuals, because they conceive of themselves as exempt from rules and conventions, are likewise unlikely to make the inference from another person's being held liable to the likelihood of liability befalling them. Other kinds of mentally ill persons are unable to understand their own motivations for action. Why they do what they do is as much a puzzle to them as it is to us; others are not even aware of what they do.

I do not want to argue that all individuals falling into these categories ought to be held liable any more than I would urge the opposite position. My point is more limited. Each of these categories of individuals (with some restrictions) might satisfy the economic standard of fault, that is, act in an inefficient manner. Nevertheless, the economic argument for imposing liability against them systematically fails either because, as a matter of personality or character, they are not individuals who would be moved by economic incentives, or because their mental deficiencies render the general deterrence argument implausible.

Here is a point at which our treatment of mentally and physically handicapped persons ought to diverge. Whereas both categories of the defective may be unable to satisfy the standard of efficiency, the physically handicapped are more likely candidates for the general deterrence argument. Even though they may lack the tools for safe and careful execution, they, unlike the cognitively deficient, have the tools for appraising conduct and its risks and for choosing accordingly. Thus, they may be encouraged by the economic incentive to forbear entirely or to curtail their activity levels to provide the margin of error their impairments require. Because the categories of mentally deficient I have been discussing lack, for example, the cognitive skills necessary to plan actions in the light of their riskiness, holding them liable is unlikely to affect their level of participation in inefficient conduct. While there may be no good economic reason for distinguishing between the mentally and physically defective in determining if the criterion of fault has been satisfied, there is a good economic reason for distinguishing between the two types of impairments in *liability* decisions.

One further point. I do not mean to contend that some mentally defective persons cannot be *prevented* from inefficient accident causing conduct. Indeed, their level of inefficiency can be lowered (where appropriate) by holding their employers liable. Imposing the burden on the employer creates for him an incentive to restrict their faulty conduct. He may do so by changing their work or, where that is costly and inefficient, by not hiring them. My point is simply that the economic argument is not suited to work on certain classes of mentally deficient persons or personality types.[8] Liability imposed on one is not likely, in other words, to be perceived by others as a good reason for restricting or curtailing entirely their conduct.

Indeed, in cases where the mentally abnomal may be at fault in economic terms, the better economic argument might support placing the burden of avoidance on their victims. That is, economic considerations could suggest imposing the costs of accidents caused by the mentally deficient on their victims, not because they are at fault, but because they are, if anyone is, more likely to be affected in a desirable manner by the economic incentive to reduce accident costs.

A theory of the role of fault in torts must provide an explanation not only of the meaning of fault but a justification for imposing liability on the basis of fault. The economic analysis of torts offers one plausible way in which the mentally ill, in spite of their deficiencies, may be at fault, but fails to justify in economic (cost-minimization) terms imposing liability upon them.

IV. COMPENSATION, FAULT AND MENTAL ABNORMALITY

The theory of moral responsibility fails to explain the sense in which the mentally deficient are at fault; the economic analysis fails to justify imposing liability upon them. What we are still looking for is a theory of fault liability that does both.

The way to begin is by distinguishing among different ways in which we characterize conduct as at fault or culpable.[9] There are occasions on which an actor is said to be 'sufficiently culpable' to be subject to liability for harms of certain kinds. For example, an infielder in a baseball game may mishandle a ground ball he is genuinely incapable of handling, but one an adequate or first-rate third baseman would have handled. He is charged with an error; and should his error prove to be a substantial contributory factor in his team's defeat, the loss may be imputed to him as his fault. He may be 'benched' and a demerit of sorts will be charged to his fielding record. Enough of such failing and he may be demoted to the minors. He may be subject to all this for his failure to satisfy a standard of adequate performance by individuals in his role, even if his failure to measure up is not his fault.

In contrast with fault judgments of the first sort, we sometimes characterize conduct as 'really' or 'genuinely' at fault. When we do so, we feel that the actor has failed to satisfy a norm of behavior fully within his grasp, e.g. the infielder who fumbles a ground ball he could have handled.

In fault judgments of the first sort, the standard contrived for the purposes of fixing liability for harms is fundamentally objective (or external). It takes almost no account of (most) personal traits or idiosyncrasies. It may therefore require more of some than they are genuinely capable of; further, protesting that fact would be to no avail in efforts to free oneself from liability to the appropriate 'penalty'.

In fault judgments of the second sort, the standard applied to each actor is subjective or internal to him. It requires no more of him than he is capable of; and there is a sense therefore in which it is necessary to explore what H. L. A. Hart refers to as the 'inner facts' to determine the actor's relative fault. An actor may therefore proffer an excuse sufficient to establish his lack of fault in this second sense, yet insufficient to force withdrawal of the weaker fault judgment.

From both objective and subjective fault judgments we can distinguish judgments of moral fault. Judgments of moral fault are distinguishable from fault judgments of the first sort by the fact that they are verifiable by applying the internal test. To that extent they resemble fault judgments of the second sort. Thus, unlike fault judgments of the first sort, ascriptions of moral fault are subject to withdrawal when sufficient evidence can establish that the actor's defective conduct is not his fault. Ascriptions of moral fault differ from all other fault judgments by the fact that the standard the act fails to satisfy must be a 'moral' one. We do not, after all, judge someone fitting of moral blame for his failings, even those which are his fault, unless they are moral shortcomings. Baseball players are not generally morally at fault for their fumbles, even if such fumblings are the result of personal carelessness, inattentiveness or negligence. Such actors may be liable to allegations of incompetence or inadequacy as well as to appropriate 'demerits' stamped on their 'records'. Ordinarily, however, these actors are not subject to expressions of moral indignation by their peers, nor do they, themselves, generally experience shame, guilt or remorse. In the end, these defects of action do not *usually* suggest a deeper moral inadequacy.

We can therefore distinguish among different ways in which fault may be used as an instrument of unfavorable appraisal and as a condition of liability. The important distinction is between fault as a standard for criticizing doings and fault as an evaluation of doers for their doings (or even for their beliefs or intentions). Because they are offered to fix moral blame, judgments of moral fault, are often judgments about the character of the accused. Where correctly alleged, they impute moral stigma to the individual and warrant appropriate feelings of indignation on behalf of his accusers, and shame, remorse, and guilt on the part of the accused. On the other hand, fault judgments of the first sort serve primarily as a vehicle for expressing an unfavorable appraisal of action; only secondarily do they imply an unfavorable evaluation of agents, and, then, only to the extent to which the doing is ultimately the doers. Because one is subject to tort liability for one's failure to satisfy standards of due care or foresight regardless of whether such failure marks a personal flaw, fault judgments in torts are allegations of defective *conduct.* They are fault judgments of the first sort; judgments of faulty action.

Those with mental abnormalities, like those with certain physical handicaps, are expected to satisfy the objective standard, and their failure to do so constitutes a fault in their *action,* but not necessarily in them. The question that remains, however, is whether we can justify imposing liability on the basis of the objective standard: That is, whether we are justified morally in focusing on the fault in the action in deciding liability rather than on the 'personal equation' and, therefore, the personal responsibility of the actor for his flawed conduct.

That liability is elsewhere imposed on the basis of objective tests of fault suggests that the *purpose(s)* for which the standard of liability is contrived plays a role in determining if an objective test comports with our notions of justice and fairness. In this light, reconsider the baseball example. Charging an error to an infielder helps us decide which runs scored are unearned, that is, which are not chargeable to the *pitcher's* record. We are deciding against whose record whatever runs may score will be charged. Charging an error to the infielder (imputing fault to him) directly benefits the pitcher whose record otherwise would be blemished. The objective test is thought to be justifiable in that context in part because liability for the runs scored must be imposed on someone, and the erring infielder is a *more suitable* candidate for it than is the pitcher.

One interesting argument to support the common law rule that one ought to be liable for one's failures to satisfy an objective standard rests upon the baseball example. In torts we are dealing with activities and their accident costs. The question is always who should bear these costs. There is, in other words, a loss – one which in a sense will not go away. Someone has to bear it. The common law approach to the problem of deciding upon the incidence of such a loss is almost always to restrict that decision to a choice between the injurer and his victim. If the choice is between a faultless victim – that is, one whose conduct fails to contribute causally to the harm or one whose conduct, though it contributes to the occurrence, in every way complies with community ideals – and a faulty injurer, one whose conduct not only contributes to the occurrence but falls below our ideals as well, the loss ought to be imposed on the party at fault.

Liability is not imposed upon the faulty injurer to penalize or to punish him. Indeed, the motivation of the argument is to resist precisely that implication. Instead, liability is presumably imposed upon the faulty injurer to protect his faultless victim from the burden of bearing his accident costs.

This argument provides one good reason for imposing liability on the basis of an objective test of fault. It suggests as well one way in which liability based on the objective standard comports with our sense of justice. Though the standard of fault in torts need not measure moral culpability, it *does not* follow that being at fault in this sense is not a *moral reason* for imposing liability in the sense of deciding upon the incidence of a loss. If a loss must fall on either of two *morally* blameless parties, in order to protect the innocence of the completely faultless individual, liability ought to be imposed on the party who has failed to comply with standard of reasonable care.

The argument is by no means conclusive. It justifies imposing liability on the basis of the objective test when otherwise a faultless victim would bear the loss. After all, the faultless victim's innocence is protected in every case in which he is not required to internalize the costs, not just in those in which his particular injurer bears them.

What we have demonstrated then is a somewhat weak thesis. It is that where there are *no other relevant differences* between injurers and victims, the fact that the injurer but not the victim is at fault ought to be decisive in imposing liability upon him. We can of course imagine other distinguishing characteristics that might constitute a moral difference between injurers and victims, e.g., their relative abilities to bear the loss or to distribute it maximally over persons and time. The point of the argument can be put this way: Failure to satisfy the objective test of fault need not mark a moral fault in an actor. Nevertheless it may constitute a good, though not necessarily conclusive, moral reason for imposing upon the defendant liability for the harmful causal upshots of his conduct.

One can read certain passages in Holmes as sympathetic to this general line of argument. In particular, after noting the difficulty a congenitally awkward individual faces in complying with our expectations of reasonable care, Holmes notes that liability is justly imposed upon him nevertheless, for his slips "are no less troublesome to his neighbors than if they sprang from guilty neglect" [11]. Protecting and enhancing the security of innocent victims is the goal of torts suggested by these remarks. Liability is justly imposed on those who cannot perform more carefully, not to punish them – for that would serve little purpose – but to guarantee that the faultless victims are not unfairly burdened.

Unfortunately Holmes elsewhere rejects just the sort of argument these anomalous remarks of his suggest. In identifying the goals of tort law, Holmes denies outright that compensating victims is among them. He reasons that if compensation is our goal, we ought to forego costly litigation and provide private or public insurance for potential victims. Similar sentiments are expressed in Posner's *Economic Analysis of Law*: "That damages are paid to victims is merely a detail" ([16], p. 78). In the end, then, both Holmes and Posner deny that compensation is a goal of tort law. On the contrary, I want to argue that only by viewing compensation as the primary goal of torts can we justify imposing liability on the basis of the objective test of fault.

The Holmes–Posner contention rests on a seriously confused assumption. It is that in torts we are concerned with compensating the general class of victims. Providing recompense to the general category of victims is simply not a goal of torts. Were it, Holmes would be right and torts would be a very expensive and cumbersome social instrument for doing so. The primary goal of torts is to identify and compensate a certain class of victims for whom compensation *is a matter of justice.*

While it is surely more expedient to render compensation either through tax coffers or through a system of first-party insurance contracts, it hardly follows that everyone thus compensated would have a moral right to it. Tort law tries to sort out from the general class of victims those who as a matter of justice are entitled to compensation. Indeed, it goes one step further. It tries to distinguish a class of victims who are entitled to recompense as a matter of *compensatory* justice from those whose claims are rooted in *other principles of justice* and *social policy.* Failure to pay heed to such distinctions underlies another of Posner's more serious confusions. He writes:

> The rejection of moral criteria as a basis for liability follows easily from the conception of fault as a compensation scheme and nothing more; it would be odd to deny welfare benefits on the ground that the recipient's misfortune was not the product of someone's wrongful conduct ([15], p. 31).

But would it be so odd? Even where we feel compensation is warranted, it may be worthwhile to distinguish among the justifications of claims to it. Requests for recompense may be rooted in considerations of general welfare, utility, wealth distribution, benevolence or justice. Different social practices may adjudicate different claims to compensation. To argue, as Posner appears to, that the compensation paradigm of torts is self-contradictory because it recognizes some but not other categories of worthy claims to compensation simply misses the point.

Torts is concerned with a narrow category of claims to recompense: in particular, a subgroup of those grounded in our principles of compensatory justice. I have argued elsewhere that compensatory justice is concerned with eliminating underserved or otherwise unjustifiable gains and losses [3]. Compensation is therefore a matter of justice because it protects a distribution of wealth – resources or entitlements to them – from distortion through unwarranted gains and losses. It does so by requiring annulment of both.[10]

On this view, tort law seeks to identify and eliminate a certain category of distortions: *those caused by the faulty conduct of another.* Where injurer fault results in unwarranted benefit to him – as in fraud – torts is concerned with eliminating that gain as well as with recifying the consequences of his conduct to others. In the absence of wrongful gain, however, the presence of fault in the injurer's conduct is evidence of the unjustifiable nature of the victim's loss. So the criterion of fault is necessary in torts either as a standard of unjustifiable gain or loss.

Conduct that falls below our standards of proper care and foresight, whether or not noncompliance marks a personal weakness in the actor, is in a suitably narrow sense undesirable or at fault. It may be, as the proponents of the economic analysis contend, that our standard of conduct is one of maximizing social utility and that noncompliance is undesirable because it is inefficient. Or it may be that the standard is one of reciprocity in risk taking and that noncompliance with it is unfair.[11] However we unpack the standard of the reasonable man, we will provide a reason why failure to live up to the standard is undesirable. Injuries caused by conduct that is at fault creates losses that are in the same sense undesirable or unjustifiable. They are not, by contrast, losses one incurs by chance, mismanagement, contractual agreement, or by an 'act' of God. They are losses one incurs because of the undesirable conduct of others: losses which, from the point of view of compensatory justice, ought to be eliminated or rectified.

We have now considered two complementary arguments for the moral justifiability of the objective test of fault; that is, for denying the propriety of 'inner' factors in assessing tort liability. According to the weaker argument, liability is justifiably imposed on the faulty injurer, whether or not the fault in his conduct marks a personal shortcoming in him, when otherwise a faultless victim would have to bear the costs. Where the victim is likewise at fault, the principle does not apply, and recovery is denied on the basis of contributory fault. Where neither plaintiff nor defendant is at fault the principle again does not apply, and in those areas of torts governed by the fault principle, the victim will bear the costs. (In areas of strict injurer liability, for example, Products Liability, the injurer will in similar circumstances be required to shoulder the costs.) In those instances, however, where there is injurer but not victim fault, as determined by the objective test, the injurer ought to bear the loss to protect the faultlessness of the victim. Though legal fault need not signify a moral shortcoming in the actor, legal fault may nevertheless constitute a moral reason for distinguishing between injurers and victims in deciding upon the incidence of a loss.

It does *not* strictly follow from these considerations that the victim is entitled to compensation; only that failure to compensate him will in some sense penalize his faultlessness and place a negative incentive on his compliance with community norms of behavior. The second argument, which derives from a general theory of fault liability, complements considerations of the first sort by suggesting that the fault of the injurer is not *just* a morally relevant distinction between victims and injurers, but where it contributes causally to the victim's injury, it suffices to ground the victim's claim to compensation as a matter of justice.

When conjoined these considerations suggest the following line of argument. That the injurer is at fault implies no more than that his conduct is in an appropriate sense undesirable. It may or may not be morally so; he may or may not be morally or personally at fault for it. Nevertheless the conduct is unjustifiable. Should his conduct contribute substantially to another's injury and consequent losses, those losses would also be in a suitably narrow sense unjustifiable. Accordingly, the victim would as a matter of justice be entitled to recovery. Given the common law tradition of deciding upon the incidence of a loss between respective injurers and victims only, it would be morally justifiable to impose liability upon the faulty injurer. A direct consequence of a failure to do so would be to create an unwarranted burden on his victims. Moreover, failure to do so would emasculate entirely the victim's claim to compensation itself rooted in principles of justice.

The argument can be further strengthened by appealing to the theory of the logical correlativity of rights and duties. Because the victim's right to recompense is a claim-right and not merely a license or privilege, it creates, on the logical correlativity thesis, a corresponding obligation to render compensation to him. In this case, the suitable candidate (barring prior contractual arrangements) would be his injurer. We can reach the same conclusion yet another way. If we conceive of torts law as identifying debts of repayment, then the right to be repaid is owed the victim and the obligation to repay falls on that party who has incurred the debt, in this case, the faulty injurer.

It does *not* follow from any of these considerations that *justice* always requires that the faulty injurer discharge the obligation of compensation or repayment to his victims. Like other debts of repayment, the obligation to provide recompense may be justly discharged by a third party. What compensatory justice does require is recompense for individuals who have absorbed unjustifiable losses. It is a feature more of the framework of the common law than of our ideals of justice that the injurer discharge that obligation personally or through his insurance plan.

To summarize: The justification for employing an objective standard of fault seems to follow not from abstract moral reasoning but from a general theory of torts. Unlike the theories of moral responsibility and loss allocation, the theory of compensatory justice is 'victim-oriented'. Liability is imposed on faulty *injurers* not to penalize their moral fault or to put a cost on their inefficiency, but to compensate individuals who have suffered losses caused by undesirable conduct. The objective standard of fault is justified therefore *not* because it provides a test of moral fault, which it does not; and *not* because liability imposed on the basis of it maximizes cost-avoidance, which, as the examples of the mentally defective and ill illustrate, it does not. The objective test of fault provides us with a tool for identifying faulty *action,* the harmful consequences of which warrant compensation. Thus, it is only by conceiving of torts primarily as a compensation scheme and fault as a appraisal primarily of acts, not actors, that the common law liability of the mentally abnormal can be squared with our conceptions of justice.

These considerations suggest important differences in the requirements of justice in tort and criminal liability. That is how it should be. In the typical tort case, Jones sues Smith alleging that Smith's failure to exercise due care caused him to suffer some loss. There is a loss that initially has fallen on Jones. All we can do is shift it to Smith or let it lie on Jones. If we decide against Smith, Jones directly benefits; he is freed from the burden of the loss. If we decide otherwise, it is Smith who benefits accordingly.

In the criminal law, no individual or group stands to benefit in a corresponding fashion. Indeed, in theory at least, deciding that the alleged offender is innocent and therefore ought not to be punished is said to benefit both him and his accusers (society). In any event, there is no criminal law analogue of the 'loss' that must be borne by someone. I do not mean to suggest that punishing an offender has no desirable social benefits. Surely, doing so may provide an opportunity for society to express its outrage and condemnatory sentiments. Doing so may also (though minutely) enhance personal security. A verdict of innocence, however, does not imply that there is an identifiable individual who must be punished, nor does society taken as a whole suffer a loss. In torts we may be searching for relevant difference between (or among) candidates to determine who ought to bear a loss. In the criminal law, we are not comparing the alleged offender's conduct with that of another as much as we are measuring it against a standard of personal responsibility.

The prevalence of insurance in torts is additional evidence of its focus on defective conduct and provides further evidence of the diverse character of the requirements of justice in torts and in criminal law. Compensation is, in almost all accident cases, transacted through insurance contracts. On the other hand, an insurance scheme for *criminal* liability would be an outrage. (Indeed, even in torts, liability insurance is not at all prevalent for *intentional*, as opposed to accidental, torts.) Criminal liability is a 'debt' to society the guilty party must himself discharge. Tort liability is a debt one can arrange to have discharged by another.

Holmes to the contrary, similar distinctions may be drawn between contract and tort law.[12] Contracts, like the criminal law and unlike torts, is concerned with one's inner facts and personal equation. The reason for delving into the inner facts is *not* (as it is in the criminal law) to assess personal culpability. Instead, it is because a contract is said to be 'a meeting of the minds'. Evidence of mental defect, emotional disturbance, cognitive incapacity and innocent mistake are obviously relevant in determining if a contract has been made, and if liability for alleged noncompliance ought to be imposed.

In no other body of law is disregard for one's personal equation so widespread and, if I am correct, so justifiable as in tort law. The question that remains is what are the consequences of the justification of the objective test of fault for the defenses of mental defect or illness? Because the objective test is all that is required to ground a victim's claim to recompense, that the defenses of mental defect or illness would serve no useful purpose in determining liability. I want to conclude this essay by arguing that such an inference is too hasty.

V. AGENCY AND EXCUSES

Though quite a good argument can be made for compensating victims of floods, earthquakes and other natural disasters, the fact is that we do not compensate them in torts. That is because only losses caused by human action fall within the province both of torts and compensatory justice; only *human action* can violate human rights.[13] Because compensatory justice and, ultimately, tort liability require human action, there may be a meaningful but as yet insufficiently explored role for mental defect or illness as a defense to tort liability.

To explain the sense in which consideration of mental abnormality may constitute an excuse, a distinction must be drawn between two kinds of excuses. First, I want to draw a preliminary distinction between justifications and excuses as defenses to liability. In offering a justification, one takes responsibility for what one has done, but tries to show that (all things considered) the act in question satisfies the standard, is required by a 'higher' standard or is an exception to the rule that such conduct is wrong. An excuse defeats liability, not by denying that the act falls below the standard or rule, but by denying responsibility for the lack of compliance.

We can distinguish further between different ways in which excuses may defeat liability. In most instances, excuses are offered to show that while in some sense the act is the actor's doing, its failure to satisfy the appropriate norm is not a reflection on his character; the failure is not, in other words, his fault. Where personal responsibility or culpability is a condition of liability, an excuse of this sort will defeat liability by forcing withdrawal of the ascription of personal fault on which it is based. For example, liability to moral blame may require that an actor's failure to comply with the requisite standard is the result of his defective character. Consequently, persuasive evidence that the failure to measure up is the result of defective skill, cognitive incapacity, mental disturbance, or imbalance could ordinarily suffice to free one from liability to moral sanction. But, because liability in torts is based on allegations of faults in acting rather than upon faults in character, evidence of this sort would not constitute an appropriate defense.

In denying responsibility for one's failure to comply with the standards of care, one might mean to deny responsibility for the act itself; that is, one may intend to deny *authorship* of it. To deny authorship or agency is to deny that, in some sense, one has acted at all; it is to deny in Hume's terms that the act "comes from within the actor". The paradigm example of a condition of this sort is that of extreme external compulsion: for example, the defendant who is pushed by a third party down a flight of stairs into the victim waiting innocently at the foot of the stairs. Here the defendant performs no independent act of his own; instead he is a useful vehicle to the fulfillment of another's intention. This is of course the extreme case, but it does provide guidelines for further categorization. The key concept is internal control. Is the actor able to form intentions and is his action the result of them? To the extent that the actor lacks capacity to form intentions or to translate them into bodily movement, he may be said to fail to act. Evidence either of incapacity to form intentions, inability to transform them into human conduct, or to understand the relation between the two *may* exonerate one, not by denying culpability, but by denying *agency*!

Tort liability is for the most part not defeasible by excuses of the first sort; that is just another way of saying that the standard of fault is an objective one. However, because considerations of compensatory justice and therefore of tort liability are based on the harms caused by faulty human action, it is altogether plausible to accept as an excuse, and as a defense to liability, evidence that the minimal conditions of agency have not been satisfied. While the excuse of mental abnormality does not and (as I have argued) ought not defeat the judgment of fault upon which liability is grounded, evidence of mental defect or illness may yet surface as a defense to liability by denying that the minimal conditions of agency are satisfied.

Just such a distinction between excuses that defeat culpability and those that defeat agency (and therefore responsibility in one sense) seems to me to have surfaced in *Weaver v Ward. Weaver*, you will recall, was the first case in which mental illness as a defense to tort liability was discussed. After holding that lunacy was not a defense to liability, the court went on to conclude that:

> ... no man shall be excused of a trespass (for this is the nature of an excuse, and not of a justification, *prout ei bene licuit*) except it may be judged *utterly without his fault*.
>
> As if a man by force take my hand and strike you, or if here the defendant had said, that the plaintiff ran across his piece when it was discharging ... so as it had appeared to the court that it had been *inevitable*.

These remarks appear irreconcilable. The case was decided under the Trespass doctrine which provided a strict liability standard of recovery, and it is not surprising therefore that lunacy as an excuse is no defense. What is surprising is that the absence of fault apparently is.

The traditional approach to reconciling these remarks has been either to ignore the brief reference to faultlessness as a defense or not to take it seriously. We can do better, if we reevaluate the *Weaver* opinion in the light of the distinction between excuses that go to the question of culpability and those that deny agency. In Trespass, lack of culpability is no defense; however, lack of agency may be. It may be in this sense that acting without blame for the court contrasts with acting utterly without fault. Indeed, the examples marshalled in to illustrate the apparent distinction seem to support this construction of the opinion.

Both examples of faultlessness as an excuse stress the lack of contribution of the actor to the sequence of events. In both cases the harm which results is *inevitable* in the sense that it is out of the defendant's control. Not himself an agent, the defendant is merely an instrument, in the first case of another person, in the second of fate, in bringing about the harm. The harm is not his fault in the sense that it is not attributable to *any act of his.*

To this point I have refrained from drawing the standard psychological distinctions between mental defect and illness. That is because the common law makes no such distinction relevant to judging if the criterion of fault has been met; and the question I have addressed in the body of the paper is whether we can justify the common law rule. Because liability in torts does not require a personal or idiosyncratic notion of fault, I have argued that we could justify the common law liability of the mentally abnormal. So long as the mentally deficient or ill are agents they can be liable in torts for their faulty actions whether or not they are morally to blame for them. The question we must now address in the light of the above discussion is whether the mentally abnormal satisfy the conditions of agency.

I do not intend to discuss every or even most categories of mental deficiency in the light of the purposes of tort liability and the theory of human agency. Such an inquiry is well beyond the scope of this paper. But I do want to close the paper by illustrating some of the ways in which the presence of mental defect or illness may defeat tort liability. In that way, I will have at least drawn attention to the kinds of factors which ought to be considered on a case-by-case basis in determining liability in torts.

The standard preliminary distinction is between mental defects and mental disorders. Classically, mental deficiency refers to a class of individuals whose scores on standard I.Q. tests of the Stanford-Binet type fall in the lower 2% of the tested population. The range is further divided into the profoundly retarded, 0–19 I.Q.; the severely retarded, 20–34 I.Q.; the moderately retarded, 35–49 I.Q.; and the mildly retarded, 50–70 I.Q.. The mean score on the Stanford-Binet test is 100, with a standard deviation of 15. Fully 75% of the mentally retarded score above 50 and are therefore only mildly or moderately disturbed.

Mental retardation is an intellectual shortcoming measured by an intelligence test. There is no necessary connection between mental deficiency and mental illness, though the incidence of pathological patterns of adjustment is greater among the mentally ill than in the general populace. As judged by the Vineland Social Maturity Scale, the severely and profoundly retarded are capable of reaching a maximum level of maturity of between three and seven years of age. As a general rule, one might argue that the severely and profoundly retarded lack the requisite knowledge about the world, causal relations and themselves to make informed judgments and to be self-motivated actors. In theory at least, their liability in torts would be questionable. The issue is unlikely to arise in practice, however, because most of the seriously retarded are institutionalized. In other cases involving the mentally defective, agency may be a matter of degree. The maximum level of maturity for the mildly retarded is twelve years of age; and at that point there is reason to believe that the individual is as much an actor as others his mental age, though less reason to believe that he will understand as well as normal twelve year olds the purpose of sanctions imposed upon him.

Mental *illness* is an organizational disorder. The mentally ill person develops a pathological process of adjustment to his environment. These processes are usually described as either neurotic or psychotic. However uncomfortable or distressing one's neurotic behavior may be, it is almost never likely to rise to the level at which it casts doubts about one's capacity for agency. Psychotic reactions might. We can distinguish between psychotic disorders of personality and character: schizophrenia would be an example of the former, sociopathology an example of the latter.

Sociopathological personalities have mental disorders which in no way appear to deprive them of the capacity for agency. Such persons have fully integrated wants, beliefs and patterns of behavior. It is just that compliance with social norms is not included within their network of wants and desires. They understand the role of sanctions in supporting a system of rules and conventions. They simply see themselves as exempt from the rules and beyond reproach. Sociopathology therefore is a fine example of a psychotic personality disorder that normally would not defeat tort liability.

Because agency requires the ability to form intentions and to act accordingly, an agent is someone with an integrated network of wants and beliefs that are translatable into conduct. To the extent that emotional disorders involve, for example, unprovoked violent outbursts as in the manic stages of manic depressiveness, agency may be lacking. Similarly, in both catatonic and paranoid schizophrenia, the disabled person may be unable to translate intentions into actions. In the case of the paranoid schizophrenic that is because he suffers from hallucinations and acts on the basis of 'foreign' beliefs. In the case of the catatonic schizophrenic, it is because his bodily movements are bizarre and apparently uncontrolled. In yet other instances, the disabled person is unaware of or unable to integrate into his conception of himself, his motivations for acting. In all such cases, an argument can be made that the individual is not the compelling force in his 'actions'. To the extent that evidence of his mental impairment sheds light on the question of his agency, it is a perfectly plausible defense to liability in torts.

To sum up: I have argued that the common law liability of the mentally defective and disturbed cannot adequately be explained on either the orthodox or the economic theories of torts. I have attempted to present instead a moral theory of torts based primarily on the principle of compensatory justice. Such a view, I have argued, justifies the common law rule. I have also tried to show that while mental defect or emotional disorder is no defense to an imputation of fault in torts, it may yet be a defense to liability by establishing the absence of agency.

University of Wisconsin-Milwaukee
Milwaukee, Wisconsin

NOTES

[1] One legal theorist who has defended in detail the distinction between the grounds of recovery and liability is George P. Fletcher. See [9].

[2] The current debate in tort theory is best characterized, I think, as one between proponents of the view that tort liability is a specific problem within the general theory of responsibility and others who argue that issues in torts are specific instances of more general problems in microeconomic theory. The most articulate and useful defense of The Responsibility Thesis is that of Richard Epstein [6]. His position is developed further in [5] and [7]. The most accessible and interesting defenses of the economic analyses are: Guido Calabresi [1] and Richard Posner [16].

[3] S. F. C. Milsom has argued forcefully that fault has always been the standard of liability, even under Trespass. See [13].

[4] For a detailed discussion of the concepts of directness and indirectness in the Trespass framework, see Charles Gregory [10].

[5] For a useful discussion of mental illness in the criminal law, see Joel Feinberg [8].

[6] For a useful introduction to the economic analysis of legal rules, see A. Michell Polinsky [14].

[7] For a collection of useful and serious discussions of the problems of mental deficiency, see Frank J. Menolascino [12].

[8] The point is not that accident costs cannot be reduced in this way: it is that the mentally deficient are unable to make the *decisions* necessary to reduce the costs on the basis of economic considerations.

[9] The following paragraphs on the distinctions between different fault judgements summarizes arguments I have made elsewhere. See [4].

[10] A distribution of resources need not itself be just for principles of compensation to warrant rectifying distortions in it. All that is required to invoke considerations of compensatory justice is for distortions in the distribution scheme to be underserved or unwarranted. Our ideals of compensatory and distributive justice may therefore diverge. Indeed, arguments from compensatory justice are 'conservative' in nature. Thus, one may even invoke compensatory ideals to protect an unjust distribution of wealth – unless, of course, the distribution itself is the result of unjust enrichments and unjustifiable losses.

[11] This is the way I would understand Fletcher's claim that fault is a standard of reciprocity [9].

[12] One ought not press the distinctions between tort and contract law in part because the contract doctrine of promissory estoppel is rooted in principles of tort liability and because the tort doctrine of absolute manufacturer's liability is almost indistinguishable from the contract doctrine of implied warranty.

[13] In this case the entitlement violated by faulty action is the right one has to the security of one's place in the distributional scheme.

BIBLIOGRAPHY

1. Calabresi, G.: 1970, *The Costs of Accidents*, Yale University Press, New Haven.
2. Coase, R.: 1960, 'The Problem of the Social Cost', *The Journal of Law and Economics* **3**, 1–44.
3. Coleman, J. L.: 1974, 'Justice and the Argument for No-Fault', *Social Theory and Practice* **3**, 161–180.
4. Coleman, J. L.: 1974, 'On the Moral Argument for the Fault System', *Journal of Philosophy* **71**, 473–490.
5. Epstein, R.: 1973, 'Pleadings and Presumptions', *University of Chicago Law Review* **40**, 556–582.
6. Epstein, R,: 1973, 'A Theory of Strict Liability', *The Journal of Legal Studies* **2**, 151–204.
7. Epstein, R.: 1975, 'Intentional Harms', *The Journal of Legal Studies* **4**, 391–442.
8. Feinberg, J.: 1970, 'What Is So Special About Mental Illness?', in *Doing and Deserving*, ed. by J. Feinberg, Princeton University Press, Princeton, Chapter 11, pp. 272–292.
9. Fletcher, G. P.: 1972, 'Fairness and Utility in Tort Theory', *Harvard Law Review* **85**, 537–573.
10. Gregory, C.: 1951, 'Trespass to Negligence to Absolute Liability', *University of Virginia Law Review* **37**, 359–397.
11. Holmes, O. W.: 1963, *The Common Law*, Little, Brown and Co., Boston, p. 86.

12. Menolascino, F. J. (ed.): 1970, *Psychiatric Approaches to Mental Retardation*, Basic Books, New York.
13. Milsom, S. F. C.: 1969, *Historical Foundations of the Common Law*, Butterworths, London.
14. Polinsky, A. M.: 1974, 'Economic Analysis as a Potentially Defective Product: A Buyers' Guide to Posner's *Economic Analysis of Law*', *Harvard Law Review* 87, 1655–1681.
15. Posner, R.: 1972, 'A Theory of Negligence', *The Journal of Legal Studies* 1, 29–96.
16. Posner, R.: 1973, *The Economic Analysis of Law*, Little, Brown and Co., Boston.

SECTION III

INVOLUNTARY CIVIL COMMITMENT OF THE MENTALLY ILL

ROLF E. SARTORIUS

PATERNALISTIC GROUNDS FOR INVOLUNTARY CIVIL COMMITMENT: A UTILITARIAN PERSPECTIVE

My purpose in this paper is to offer an interpretation of what I believe to be the central argument underlying the strongly antipaternalistic position taken by the utilitarian moral philosopher John Stuart Mill in his classic essay *On Liberty*, and to explore the implications of that argument for ***paternalistic*** justifications of the involuntary civil commitment of the mentally ill. Although much of what I shall say will pretty obviously apply to non-paternalistic justifications of involuntary commitment as well (e.g., posing a threat of harm or gross nuisance to others), my central concern is with the morality of the laws and institutional practices reflected in statements of the following sort: From the American Psychiatric Association's revised position statement on involuntary hospitalization of the mentally ill:

> The American Psychiatric Association is convinced that most persons who need hospitalization for mental illness can be and should be informally and voluntarily admitted to hospitals in the same manner that hospitalization is afforded for any other illness . . . .
>
> Unfortunately, a small percentage of patients who need hospitalization are unable, because of their mental illness, to make a free and informed decision to hospitalize themselves. Their need for and right to treatment in a hospital cannot be ignored. In addition, public policy demands that some form of involuntary hospitalization be available for those mentally ill patients who constitute a danger . . . to themselves . . . .[1]

From the Washington, D.C. Hospitalization of the Mentally Ill Act:

> (involuntary hospitalization is authorized if a person) is mentally ill, and because of that illness, is likely to injure himself . . . if allowed to remain at liberty.[2]

The operative notions in these statements – "in need of treatment", "dangerous (or potentially harmful) to oneself" – are hopelessly vague and may be so broadly construed as to be virtually coextensive with the equally vague concept of mental illness itself.[3] They may embrace everything from the risk of suicide at one extreme to being maladjusted and presumably standing to benefit from the benevolent ministrations of a mental health care professional at the other. Must we, then, in order to examine paternalistic justifications for civil commitment, consider all of the points on the spectrum suggested? Or should we try to draw a line – in terms of the magnitude and likelihood of the harm in question, perhaps – on one side of which would fall

B. A. Brody and H. Tristram Engelhardt, Jr. (eds.), Mental Illness: Law and Public Policy. 137–145.

those cases in which involuntary commitment was justified and on the other side of which it was not? Neither of these potentially tortuous routes need be taken, of course, unless one believes that there are some cases in which involuntary commitment is justifiable on paternalistic grounds. It is precisely this belief which, following Mill, I intend to oppose. Before turning to Mill's argument, though, some further preliminary remarks are in order.

In 1968 Alan Dershowitz estimated that there were some one million persons locked behind the doors of state mental hospitals [2]. No discussion of involuntary commitment of the mentally ill would be complete without noting that they most likely represent merely a small proportion of those who are committed against their will and who would not be were it not for the existence of statutes authorizing involuntary commitment. For many of those who 'voluntarily' commit themselves to mental institutions do so on the basis of the often well founded belief that if they do not commit themselves voluntarily they will be committed involuntarily, with the added trauma of the procedures necessitated by formal commitment proceedings. How many persons fall into this category is almost impossible to estimate, but I am sure that the number is substantial. Any assessment of the social consequences of involuntary civil commitment as an institutional practice must surely take such persons, and their undesired loss of liberty, into account.

As the preceding remarks suggest, it is the justifiability of a set of institutional practices defined by law and given life by the customary behaviors of judges, attorneys, health care professionals and others that is at issue. Few of us would be at a loss to describe the details of a particular case, real or hypothetical, where we would have no reservations about the involuntary commitment of one mentally ill as a means of protecting that person against himself *within the framework of existing law and institutional practice*. But this says absolutely nothing about the justifiability of the practices themselves. Where acts of a certain kind are legally permitted, it may be that people on occasion ought to perform them. But it may also be the case that they ought not to be permitted in the first place.

It is worth noting that neither the American Psychiatric Association's Statement nor the D.C. Hospitalization Act provide for the involuntary commitment of those who are simply (a) mentally ill, and (b) dangerous to themselves. The D.C. Act requires that the latter (dangerousness) be 'because of' the former, while the APA Statement requires that 'because of' mental illness the patient be unable "to make a free and informed decision" to hospitalize himself. Indeed, on one reading the APA Statement does not require dangerousness to self (or others) at all, but in fact proposes two independent paternalistic grounds for 'involuntary commitment': (1) mental illness plus dangerousness to self; (2) mental illness in virtue of which the patient is incapable of making a rational decision as to whether or not to commit himself.[4] For the sake of the argument, I would like to consider a proposal which may thus be more restrictive than either that of the APA or the D.C. Act. It would authorize involuntary commitment (on paternalistic grounds) only where the patient was (a) mentally ill, (b) dangerous to himself in some quite clear and uncontroversial sense, (c) incapable of making a rational decision to commit himself, and (d) both (b) and (c) were due to (a).[5]

I have framed the above criterion for involuntary commitment on the grounds of dangerousness to self with a view toward going at least some way toward meeting the frequently heard objection that justifying interferences with individual liberty on such grounds would surely justify too much. For there are many activities which pose considerable danger only to those voluntarily choosing to engage in them and which we would yet feel it illegitimate to interfere with on the basis of paternalistic concerns. But how are we to distinguish the mentally ill who are dangerous to themselves from sky divers, mountain climbers, heavy drinkers and smokers, and those who refuse available treatment for serious physical disorders? The difference, where there is one, lies in the capacity of the individual to make a rational choice to change his ways. And even where such a capacity may be lacking – as in the case of the individual who is just too stupid to appreciate the dangerousness of his activities – it may not stand in the appropriate causal relationship to the dangerous activity. But where it does, a criterion analogous to that framed for mental illness can surely be satisfied, and when it is, paternalistic interferences with individual liberty are surely as justifiable as they are in the case of the mentally ill. Drug addiction and alcoholism are clear cases in point. Indeed, given the argument which I shall shortly develop, it will turn out that paternalistic interference with individual liberty may be justified in some such cases but not in the case of those who are dangerous to themselves by virtue of being mentally ill. At any rate, what I am suggesting is that those who are in favor of paternalistic treatment of the mentally ill must admit to the acceptance of a quite general criterion of the following sort: If a person is in some condition C in virtue of which he is both dangerous to himself and incapable of making a rational decision to adopt patterns of thought and action which will render him less dangerous, then the state may legitimately interfere with his liberty as a means of preventing him from harming himself. Such a principle does not require that the state have and make available means of ameliorating C (treatment in the case of mental illness) but it should be understood to require that the disvalue associated with the loss of liberty in question be outweighed (for a rational agent) by the benefit attached to the avoidance of the potential harm (this to be calculated in terms of both its magnitude and its likelihood of occurrence).

Mill's *On Liberty* [7] is unequivocal in its opposition to any such paternalistic position. It is devoted, writes Mill, to the defense of :

> one very simple principle, as entitled to govern absolutely the dealings of society with the individual in the way of compulsion and control .... That principle is that the sole end for which mankind are warranted, individually or collectively, in interfering with the liberty of action of any of their number is self-protection. That the only purpose for which power can be rightfully exercised over any member of a civilized community, against his will, is to prevent harm to others. His own good, either physical or moral, is not a sufficient warrant ([7], p. 135).

Mill, as an act-utilitarian, is committed to the overarching moral principle that the morality of any particular act is to be determined by an evaluation of *its* consequences – in terms of the production of satisfaction and dissatisfaction – for all of those who will be affected by it. How, then, could he consistently propose an *absolute* prohibition upon anything of real interest, let alone paternalistic legislation or particular acts motivated by paternalistic considerations? In attacking Mill's principle in his *Liberty, Equality, and Fraternity* in 1882, James Fitzjames Stephen contended that this question simply could not be consistently answered.[6] His argument is that no matter how great the value of human liberty, the utilitarian is committed by the very nature of his position to deciding each case (of possibly justified paternalism) on its individual merits.

> If ... the object aimed at is good, if the compulsion employed is such as to attain it, and if the good obtained overbalances the inconvenience of the compulsion itself, I do not understand how, upon utilitarian principles, the compulsion can be bad.[7]

Stephens's argument has been accepted by many recent commentators on Mill's position. Gerald Dworkin, for instance, writes that "A consistent Utilitarian can only argue against paternalism on the grounds that it (as a matter of fact) does not maximize the good. It is always a contingent question that may be refuted by the evidence" ([3], p. 193). Mill's utilitarian arguments against paternalism, Dworkin contends, must be understood as being accompanied by a second line of (quite non-utilitarian) argument which places an absolute value upon the autonomy of human choice ([3], pp. 193–194). Joel Feinberg has explicitly endorsed Dworkin's account, claiming that the argument in favor of an absolute prohibition upon paternalism must rest upon the recognition of an 'abstract right' to freedom of choice the violation of which would be "an injustice, a wrong, a violation of the private sanctuary which is every person's self ... whatever the calculus of harms and benefits might show" ([4], pp. 108–109).

Let us look, then, at the text of *On Liberty* with a view to determining whether or not Mill's arguments are sufficient to support his absolute stance against paternalism without being supplemented with a non-utilitarian appeal to an abstract right of self-determination.

> But the strongest of all the arguments against the interference of the public with purely personal conduct is that, when it does interfere, the odds are that it interferes wrongly and in the wrong place. On questions of social morality, of duty to others, the opinion of the public, that is, of an overruling majority, though often wrong, is likely to be still oftener right, because on such questions they are only required to judge of their own interests, of the manner in which some made of conduct, if allowed to be practiced, would affect themselves. But the opinion of a similar majority, imposed as a law on the minority, on questions of selfregarding conduct is quite as likely to be wrong as right, for in these cases public opinion means, at the best, some people's opinion of what is good or bad for other people ... ([7], pp. 214–215).

> The interferences of society to overrule his judgment and purposes in what only regards himself must be grounded on general presumptions; which may be altogether wrong, and even if right, are as likely as not to be misapplied to individual cases .... All errors which the indvidual is likely to commit against advice and warning are far outweighed by the evil of allowing others to constrain him to what they deem his good ([7], p. 207).

While Dworkin describes it as 'fairly weak' ([3], p. 194) and Feinberg as 'at best strong' ([4], p. 108), they are one in agreeing that "arguments of this merely statistical kind ... (only) create a ... rebuttable presumption against coercion of a man in his own interest" ([4], p. 108). Both writers, I believe, miss the thrust of Mill's arguments. They are correct in claiming that they establish only a presumption in any particular case, but fail to see that they are sufficient (assuming that they have a sound empirical foundation) to establish an absolute prohibition upon paternalism at the level of law and institutional practice. If the considerations which Mill relies upon are applied not to the question "Do we coerce Jones for his own good?" but to the question "Do we permit legal authorities to ever coerce anyone for their own good?" it is obvious that it is an absolute prohibition of a certain kind of conduct rather than a rebuttable presumption against the permissability of particular acts of that kind that is thereby established.

The argument which Mill described as "the strongest of all arguments" in favor of his anti-paternalistic principle is a powerful one, and would appear to be the only kind of argument to which the utilitarian could consistently appeal in an attempt to give Mill's principle the absolute status that he quite rightly required for it. In addition, it would seem to be the only kind of argument which is not damaged by the admission that there are specific instances in which particular acts would be justified on paternalistic grounds if they were not antecedently generally prohibited. The general schema of the argument, I suggest, is the following. Assume (1) that most acts of kind K are, on utilitarian grounds, wrong, although (2) some acts of kind K are, on utilitarian grounds, right, but that (3) most attempts to identify exceptions to the rule of thumb "Acts of kind K are wrong" are mistaken because there is no reliable criterion by means of which exceptions to the rule may be identified. Where these conditions are all satisfied, the act utilitarian has good reason, other things being equal, for acting so as to prevent anyone from ever performing an act of kind K. Whereas (1) and (2) by themselves establish only a rebuttable presumption as to how one ought to act in particular cases, the addition of (3) provides the grounds for an absolute prohibition on the kind of activity in question.

One would surely be inclined to argue this way on the Constitutional level with respect to First Amendment freedoms. (1) Most governmental attempts at interferences with, e.g., freedom of the press, have had bad consequences and are thus wrong on utilitarian grounds. (2) But one can of course think of specific instances in which legal interference with a particular publication would have good consequences. (3) On the other hand, were government to have the legal power to decide that something was an exception to the hands-off policy indicated by (1), more often than not decisions to interfere would be mistaken, with bad consequences in specific cases, and a 'chilling effect' on the press in general stemming from well-founded fears that the power would be abused. Therefore, an absolute prohibition on legal interference with the press is the policy choice which will have the best consequences in the long run. Although this may not be very elegant as a piece of Constitutional analysis, it should suffice to remind us of what is a familiar and totally reasonable pattern of argument which is wholly consistent with an act-utilitarian position. That the *act* in question is in this case one of the choice of a constitutional *rule* makes it no less an act. I therefore contend that Mill's argument is a valid one; that the conclusion which he reaches does follow from the premises which he puts forward in support of it, that conclusion being that "mankind are greater gainers (in the long run) by suffering each other to live as seems good to themselves, than by compelling each other to live as seems good to the rest" ([7], p. 138).

Although valid, the soundness of Mill's argument in any particular application will depend upon the truth of the empirical assumptions embodied in the corresponding instantiation of the argument schema outlined above. Surely some forms of paternalism are justified – Mill himself admitted exceptions to which most of us would be willing to add.[8] But although his objections to paternalism may be overdrawn (although I do not think by very much), Mill's argument does force us to address the central issue where any particular form of paternalism is involved. Namely: What are the likely long-run consequences in terms of human well being of the adoption of the laws, policies, and institutional practices which provide the framework without which it is (typically) impossible to raise the question of the proper disposition of particular cases?

What, then, are the implications of Mill's argument with respect to the involuntary civil commitment of the mentally ill on the grounds that they are dangerous to themselves? Recall that for the sake of argument we are considering a quite stringent criterion, one that would authorize involuntary commitment only where the individual was found to be not only mentally ill and dangerous, but also incapable of making a rational decision to commit himself, and it was also found to be the case that the dangerousness to self and the requisite incapacity were caused by the mental illness. The acceptability of any such criterion must hinge both upon the reliability with which it can be applied and the consequences of applying it to a population of a certain assumed composition. *Where the characteristic to be identified is relatively rare, and the costs of its misidentification at all appreciable, virtually any criterion short of one which is not 100% effective will be clearly unacceptable, even assuming that the benefits associated with the identifications it correctly makes are very considerable.* Modifying only slightly a hypothetical example found in the classic paper by Livermore, Malmquist, and Meehl ([6], p. 84), we may assume that there is a certain form of mental illness such that virtually all of those who have it will commit suicide if left at liberty, and that the other conditions imposed by our proposed criterion are satisfied as well. Again, assume that one person in a thousand suffers from the illness in question, and that our criterion is reliable, distinguishing with 95% effectiveness between those who will commit suicide in virtue of the illness in question from those who will not. In a population of 100 000, 95 of the 100 who would commit suicide would be identified and presumably benefited by being involuntarily committed. Five persons would go undetected and – unless they met a natural death first – commit suicide. But out of the 99 900 people who would not commit suicide, 4995 would also be identified and committed as suicidal!

Is the involuntary civil commitment of the mentally ill on the grounds of dangerousness to self morally justifiable? The question posed by Livermore, Malmquist, and Meehl surely answers itself: "If, in the criminal law, it is better that ten guilty men go free than that one innocent man suffer, how can we say in the civil commitment area that it is better that fifty-four harmless people be incarcerated lest one dangerous man be free?" ([6], p. 84).

University of Minnesota
Minneapolis, Minnesota

NOTES

[1] See American Psychiatric Association [1]; reprinted in Gorovitz ([5], p. 181).

[2] Quoted in Dworkin [3]; reprinted in Gorovitz ([5], p. 187).

[3] See Dershowitz [2] and Livermore [6]. My debt to both these papers, especially the latter, is substantial.

[4] This way of putting the matter suggests a possibility which the authors of the APA Statement evidently did not consider: that of a person who because of his mental illness makes an irrational decision to commit himself.

[5] As Dershowitz [2] urges, not all of those who are mentally ill are incapable of making an informed decision as to whether or not to commit themselves.

[6] Parts of the following discussion are taken from my article [9].

[7] Stephen [10]; reprinted in part in [8].

[8] Mill recognized children and 'the uncivilized'. Dworkin ([3], pp. 186–187) lists a number of plausible candidates.

BIBLIOGRAPHY

1. American Psychiatric Association: 1973, 'Position Statement on Involuntary Hospitalization of the Mentally Ill (Revised)', *The American Journal of Psychiatry* **130** (3), 392.
2. Dershowitz, A. M.: 1968, 'Psychiatry in the Legal Process: A Knife That Cuts Both Ways', *Judicature* **51**, 370–377.
3. Dworkin, G.: 1972, 'Paternalism', *The Monist* 56 (1), 64–84.
4. Feinberg, J.: 1971, 'Legal Paternalism', *Canadian Journal of Philosophy* **1** (1), 105–124.
5. Gorovitz, S. *et al*: 1976, *Moral Problems in Medicine*, Prentice-Hall, Englewood Cliffs, New Jersey.
6. Livermore, P., Malmquist, P., and Meehl, E.: 1968, 'On the Justifications for Civil Commitment', *University of Pennsylvania Law Review* 117 (1), 75–96.
7. Mill, J. S.: 1962, 'On Liberty', in Mary Warnock (ed.), *Mill: Utilitarianism and Other Writings*, World Publishing, New York.

8. Radcliff, P.: 1966, *Limits of Liberty: Studies of Mill's 'On Liberty'*, Wadsworth Publishing Co., Belmont, California.
9. Sartorius, R. E.: 1972, 'The Enforcement of Morality', *Yale Law Journal* **81** (5), 891–910.
10. Stephen, J. F.: 1882, *Liberty, Equality, Fraternity,* Henry Holt and Co., New York.

DAN W. BROCK

INVOLUNTARY CIVIL COMMITMENT: THE MORAL ISSUES

In considering some of the distinctly moral issues raised by the treatment of the mentally ill, perhaps the most striking fact is the frequency with which coercion in one form or another is used on the patient, the treatment being, as a result, involuntary. Coercion is most obvious and dramatic when persons are involuntarily committed to mental hospitals, though it of course occurs in more subtle ways as well. I shall therefore discuss a number of related moral issues that this use of coercion creates, with particular emphasis on the justification of involuntary civil commitment; the principles relevant to involuntary commitment are relevant as well to involuntary treatment of committed or uncommitted persons. My aim will be to lay out some of the underlying moral principles and distinctions on the basis of which such issues can be profitably discussed and clarified, even if not settled. Since involuntary commitment of the mentally ill has frequently come under attack, I will be particularly interested in which, if any, of the common arguments against it are successful. While laws governing civil commitment provide an obvious point of departure for much discussion in this area, I will be concerned here with the moral positions that ought to form the basis of legal policy.

Since I treat a number of issues that overlap somewhat, it may be helpful to provide a brief outline of the argument's structure. In Section I, I discuss justifications for penal commitment in the criminal justice system in order to evaluate some attacks on the involuntary commitment of the mentally ill based on certain striking differences in the civil and penal system. In the second section, I consider which of the justifications for penal commitment are relevant as well to the involuntary commitment of several different kinds of mental patients, and detail what seems to me the most serious objection to commitment of the mentally ill on grounds of dangerousness. Section III is devoted to contrasting two different sorts of moral theories that underlie much of the moral disagreement in this area, and in particular that concerning the moral difference, if any, in prevention of harm to self as opposed to harm to others as a basis for involuntary commitment. In Section IV, I turn to a discussion of paternalism and some of the different sorts of failures of will and reason that might justify involuntary commitment of a person for his own good. Finally, in the last section I consider what role the concept of

B. A. Brody and H. Tristram Engelhardt, Jr. (eds.), Mental Illness: Law and Public Policy. 147–173.

mental illness should properly play in moral and legal decisions about involuntary commitment.

I

A helpful method for shedding some light on the moral justification of involuntary civil commitment of the mentally ill on grounds of dangerousness is to contrast civil commitment with penal commitment, and the moral justifications commonly offered for it. Here, philosophers have the benefit of a fairly considerable literature and a long-standing concern with the justification of criminal punishment on which to call. In considering the system of criminal law we need to distinguish two separate questions at the outset. First, what is the general purpose or reason for having a system of criminal law? And second, what is the purpose of criminal punishment, specifically the imprisonment of those found guilty of violating the criminal law? The usual answer to the first question is that the purpose of the criminal law is to reduce the incidence of certain unwanted actions. While this is not incorrect, it is an incomplete account of the purpose of the criminal law that leads in turn to an incomplete account of the justification for criminal punishment. It suggests the utilitarian bases of criminal punishment, but leaves out the non-utilitarian bases necessary for an adequate explanation of some of the central differences in penal and civil commitment. Let me briefly enumerate the usual justifications offered for criminal punishment in order to see whether they also apply to involuntary civil commitment, and where they do not, what differences in the involuntary civil commitment process are required to reflect this.

1. Deterrence

The threat of punishment for performing a criminal act is designed to deter persons who might otherwise be tempted to perform the act. The actual execution of punishment is required only when the threat of punishment has failed in that instance to deter. The deterrent effect assumes that people's choices can be affected by changing the consequences of those choices, and operates successfully to the extent that the use and threat of criminal punishment causes persons not to violate the law when they otherwise would have – either the punished person himself in his future behavior or, and more often, other non-violators of the law. In so far as deterrence is the purpose of criminal punishment, such punishment treats the criminal as a means principally to affect the behavior of others.

2. *Protection*

Criminal punishment, at least in the form of imprisonment, also has the purpose of protecting society, specifically other members of society, from dangerous persons and the harms they create. In the criminal law, protection can only be provided against persons convicted of past criminal acts; dangerousness, in so far as it justifies protection by imprisonment, is not a predictive category in the criminal law, as it is in civil commitment laws. In addition, the criminal law proscribes well-defined classes of acts such as homicide and fraud, whose meanings are in turn spelled out in considerable detail in statute and case law, and not simply 'dangerous acts' or 'acts harmful to others', categories commonly used in civil commitment laws. The purpose of protection served by criminal punishment is importantly limited by other principles of the criminal law, since prisoners cannot be held beyond the terms of their sentence, however dangerous they may still be, and cannot be incapacitated in order to make them non-dangerous.[1] Dangerousness with its related goal of protection is never a sufficient ground for initial imprisonment (a criminal act is required) nor for continued imprisonment beyond the term of the sentence.

3. *Rehabilitation*

Rehabilitation or reform of the convicted criminal is designed to change his character, motivations, opportunities, skills, and so forth, so that he will not violate the law again upon his release from prison. The forms it may take are limited by rights an imprisoned criminal retains. In general, rehabilitation is designed to *improve* the prisoner and his situation, not simply to make him non-dangerous, which latter could be achieved through various kinds of massive incapacitation, the use of which would outrage most of us. The evidence indicates that our present prison system does very poorly in terms of rehabilitation, and prison authorities are becoming increasingly cognizant of the view that available methods of rehabilitation are ineffective with respect to a substantial proportion of the prison population. A similar limitation in the effectiveness of available treatment techniques affects mental health professionals in their treatment of much mental illness.

These first three purposes, or justifications, of criminal punishment fit the general utilitarian model of the criminal law as a method of reducing the incidence of certain unwanted actions: deterrence, as it acts principally on others besides the imprisoned criminal and makes them less likely to perform similar illegal actions; protection, as it prevents the convicted criminal while imprisoned from performing additional acts harmful to others (or at least to others outside the prison); and rehabilitation, as it makes the convicted criminal himself less likely to perform further criminal acts once he is released from prison. In any particular instance of criminal punishment, not all of these purposes need be realized; for example, protection may occur during the term of imprisonment in cases where rehabilitation is an utter failure. Both for the justification of the overall practice of criminal punishment, as well as for any particular instance of it, the absence of any one of these preventive justifications is not sufficient to show that justification is entirely lacking.

It is when the account of the justification of penal commitment stops with these first three purposes that certain striking differences in the penal and involuntary civil commitment processes seem especially troubling. In particular, two arguments against the involuntary commitment of the dangerous mentally ill are commonly made. First, it is a commonplace of Anglo-American jurisprudence that it is better for ten guilty persons to go unpunished than for one innocent person to be punished. And this is not simply given lip service and forgotten, but is given expression in various rules of evidence and procedure within the criminal justice system designed to insure that when we fail to convict all and only the guilty, as we inevitably will, we fail to convict the guilty far more often than we mistakenly convict the innocent. Available evidence suggests, however, that in the civil commitment process our toleration of error is exactly reversed. Studies indicate that when psychiatrists find mentally ill persons to be dangerous, or likely to cause substantial harm to themselves or others, they tend to greatly over-predict dangerousness [3]. For each failure to correctly predict as dangerous a person who subsequently commits an act causing serious harm to himself or others, numerous persons will be judged dangerous who do not in fact turn out to be such; in statistical terms, the false positives are substantially greater than the false negatives. This precisely reverses the toleration of error found in the criminal justice process. Why do we prefer to let the guilty go free rather than wrongly convict the innocent under the criminal law, while preferring in dealing with the mentally ill to commit involuntarily the harmless rather than let a dangerous person go free? Can there be any justification for this strikingly increased willingness wrongfully to confine the mentally ill than the convicted criminal?

The second important difference leads to a more wholesale attack on the involuntary commitment of the mentally ill for dangerousness. With very limited exceptions, preventive detention – the detaining of persons who either have committed no offense or have not yet been convicted of committing an offense – is not accepted in the criminal justice system. The government occasionally attempts to claim or exercise such a power, and limited use of preventive detention is sometimes defended, but it is another commonly accepted principle of Anglo-American criminal law that a person is not to be imprisoned and so to lose his liberty until he has been convicted of a violation of the criminal law.[2] Here again, the contrast with procedures allowing involuntary commitment of the mentally ill on grounds of dangerousness is striking. *All* involuntary commitment based on a prediction that a person is dangerous, that is, will at some point in the future commit an act causing serious harm to himself or others, is preventive detention. Even in states where such predictions must be based in part on past harmful acts, involuntary commitment is still preventive detention since the commitment is not for past acts, but rather the past acts are intended only to serve as strong evidence of the likelihood of future harmful acts. But if preventive detention is abhorrent in the criminal law process, why do we allow it as a commonplace, as the *standard procedure*, in the involuntary commitment of the mentally ill on grounds of dangerousness? Can there be any justification for denying the mentally ill this important protection afforded criminals?

There is a fourth justificatory element of criminal punishment which provides a partial explanation of these differences, and sheds light in turn on the justification of the involuntary commitment of the mentally ill. It is the justification from justice or fairness. The legal system can be seen, at least ideally or when it is just, as setting up a fair or just adjudication of the competing claims that citizens make on each other and on their common social institutions; it sets up a distribution of benefits and burdens, rights and duties or responsibilities.[3] In particular, the criminal law requires everyone to forbear from certain specified acts that cause harm to others, and results, so far as the law is obeyed, in everyone being free from having those harms caused to themselves. The benefit is being free from specified harms such as being assaulted or killed; the burden is the loss of liberty to act in ways that cause such harms to others when it may be in one's interest to do so. If the body of criminal law is itself justified, then there is an overall gain in benefits obtained over burdens imposed. The criminal then acts unfairly in receiving the benefits of others' forbearance from harming him while failing to forbear from harming others; he benefits from the increase in security of person all obtain under a system of criminal law, while failing to do his part in the burdensome activity necessary to produce that benefit. Criminal punishment, therefore, has among its purposes the aim of responding to and righting this unfairness; by imposing the burden of imprisonment, it helps to restore the just balance of benefits and burdens that the criminal law ought to establish, and prevents the criminal from profiting from his wrongdoing. This gives at least part of the basis for the principle that violators of the criminal law *deserve* punishment, one form of which may be loss of liberty by imprisonment, a principle the three earlier utilitarian sorts of justification fail to account for adequately. Punishment is *required* by justice or fairness, and need not simply be revenge even if there were no utilitarian sort of justification.

Another aspect of the justice or fairness justification emerges with the conception (made well known by H. L. A. Hart) of the law as a choosing system ([12, 13]). The criminal law can be understood as warning us to refrain from proscribed acts such as homicide, or suffer punishment. The excuses that in the criminal law remove a law-breaker's liability to punishment – roughly, coercion, inability to conform to the law, and nonculpable ignorance of the nature of one's action – are designed to insure that the violation of the law was intentional and voluntary. They are designed to prevent violation of the moral principle that no one should be punished who could not reasonably have helped doing what he did. This conception of the law as a choosing system helps explain the basis for the principle that *only* the guilty should be punished, that only law-breakers *deserve* punishment. If the law in effect says, abstain from proscribed acts or be punished, then only if one satisfies the antecedent condition of being a lawbreaker does one deserve the consequent of punishment. Since the law implicitly promises that if one does refrain from performing proscribed acts, one will be free from punishment, it would be a serious unfairness or injustice to punish those not found to have been lawbreakers. This aspect of the justification of criminal punishment from justice or fairness also makes clear that in this conception of the law the citizen is viewed as a rational chooser or decision-maker, who is held responsible and accountable for his choices not when they are in any sense uncaused, but rather when they are knowing and uncoerced.

The two striking differences in our treatment of criminals and the mentally ill should now begin to make more sense. Because it should be possible to say of the criminal that he has acted freely, knowingly, and unfairly in performing an expressly and publicly forbidden act, and in the case of most of the criminal law, performed an act in itself morally wrong (homicide, assault, fraud, etc.) apart from the charge of unfairness, there is an expressive function to a finding of guilt in the criminal process ([7], Ch. 5). Specifically, there is a charge that he has performed a forbidden act and taken an unfair advantage for himself, and that given the absence of one of the usual exculpatory excuses, his action was intentional and he could have helped doing what he did.

There is in the institutional process of the criminal justice system, a public expression of disapproval and condemnation of the individual for his action and for his character. It is this condemnation, as well as the loss of liberty present in both civil and penal commitment, that the bias towards not punishing the innocent is designed to avoid, for in the case of the innocent such condemnation is misdirected. Because of this false labelling of the person and his character when an innocent person is punished in the criminal law system, there is an additional wrong done to him, and so we have an additional reason to avoid wrongly punishing him which is not present in the case of civil commitment. Involuntary civil commitment of the mentally ill is not punishment in this sense. It is not punishment, even when the commitment is made on grounds of dangerousness, for an action for which one could be held responsible. Therefore, the public condemnation essential to punishment is absent.

One could complain that while the civil commitment process does not result in an expression of condemnation as does the criminal justice system, nevertheless it does result in a stigmatization of the mental patient that is equally if not more injurious to him. So we have just as much reason to avoid error in findings of dangerousness that will lead to civil commitment as in findings of guilt. I find it quite difficult to weigh the relative seriousness of the condemnation of criminals against the significantly different stigmatization of mental patients, but certainly in the sense that there is a serious harm done in each case, this objection is sound. However, there remains an important difference between the condemnation of criminals and the stigmatization of mental patients besides the different sorts of negative judgments directed to the criminal and to the mental patient. In the former instance, the condemnation is intended and is not based on misinformation or prejudice; it is an integral part of the system of criminal law and could not be removed without radically changing our view of that system. The stigmatization of committed mental patients, on the other hand, though serious, is incidental to that process. It is to a large extent a result of public prejudices and false beliefs about mental illness. And that means we have the alternative of attempting to change and remove these beliefs and prejudices in order to remove the stigmatization that they produce. That seems to me one of the most pressing tasks required by the problem of stigmatization, but it is not an alternative we have in the criminal law system. I would emphasize, however, that so long as stigmatization of mental patients persists, we have perhaps as strong a reason to attempt to avoid incorrectly labelling persons as mentally ill as we do unjustly labelling persons as criminals, and so we do not have justification for the significant reversal of toleration of error found in the civil in contrast to the penal process. Given the important difference in civil and penal commitment disclosed in the justice or fairness justification, and the different sort of labelling in the two processes, an attack on the civil commitment process based solely on its contrast with the protection of the innocent in the criminal process is unsound. While both processes involve loss of liberty to the person committed, they are too different to allow any straightforward argument from the nature of this aspect of the penal process to the desirable nature of the civil commitment process.

The second contrast, the use of preventive detention in civil commitment, should now make a bit more sense as well. Given the 'forbear or be punished' model of law, only those who have been found to be violators can be punished fairly. Again, it is this unfairness as well as the loss of liberty that is objectionable in preventive detention; preventive detention is incompatible with this model of the law as a choosing system. The law is designed to allow persons to control and predict when they will encounter sanctions such as fines and imprisonment by making their imposition follow only upon acts known beforehand to have such consequences. Preventive detention breaks this link between choice and sanction, and so destroys our ability to predict and control when we will be subject to such sanctions. Involuntary civil commitment is detention in order to prevent acts harmful to the detained or others, and in this sense is preventive detention. But it is not unfair in the same way that it is in the criminal law system because there is no analogous choosing system in the civil commitment process for preventive detention to be incompatible with. Nor is involuntary civil commitment for dangerousness, when no harmful act has yet been performed, always punishment of the innocent in the same sense as in the criminal process. Since the condemnation inherent in the penal process is lacking, there is no claim made of the innocent person being guilty; there is no punitive element, no punishment, and thus no attribution of guilt or innocence.

While preventive detention of the dangerous mentally ill may thus lack one important objectionable feature that such detention has in the criminal law, we need to consider further why we are justified in abandoning, when we deal with the mentally ill, the restriction observed in criminal law whereby only those who have already committed a criminal act may have their liberty taken away. Why is a preventive approach to harmful acts by means of involuntary detention justified with the mentally ill when it is not considered as such for other citizens? Dangerous persons can often be identified in the general population, and certainly among past criminal offenders, as reliably as among the mentally ill, but they cannot be committed for purposes of protection. The difference therefore cannot be that protection is possible in one instance but not in the other.

We can use the law to reduce the incidence of specified unwanted acts because adults in general have the capacity to guide their behavior in accordance with promulgated social rules. It is roughly the capacities related to the notion of autonomy – being able to form purposes, weigh alternatives according to how they fulfill those purposes, and act on the result of that decision process – that make it possible to redirect human behavior with promulgated social rules. The criminal sanctions attached to specified acts are designed to make those acts less attractive in the deliberative process. Our use of this method of controlling criminal behavior, which may at least in some kinds of cases be less effective than a predictive/preventive approach, reflects several things: first, the important value ascribed to individual autonomy, to persons controlling, being responsible for, and held accountable for their own behavior; second, the importance given to the individual's rights of liberty and privacy, both in our moral thinking and our constitutional tradition; third, the great potential for error and abuse present in a system that would allow any group of persons, whether government authorities or physicians, to deny other persons their liberty based only on a prediction that they might at some point commit criminal acts.

The dangerous mentally ill are treated differently, I believe, at least to the extent that the difference purports to be justified, because they are not considered capable of guiding their behavior by promulgated social rules. As a result, they are not subject to the 'do it or else' criminal law, but rather to a cost/benefit calculus of protection and prevention. To put the point dramatically, we lock them up as we would a rabid dog (though our prediction of dangerousness in this case is a good deal more reliable), because their behavior cannot be guided and controlled by the more desirable means of laws. If this is the underlying assumption, then a far more careful sorting out of the different forms of mental illness, and of mentally ill persons, is required with regard to the effects of those illnesses on a person's capacity to understand and follow legal rules. Much that is now classified as mental illness does not seem to substantially impair that capacity, nor to impair it below some reasonable minimum level; the capacity of many hospitalized mental patients to follow hospital rules and regulations suggests this. Just when these capacities are seriously impaired is a difficult empirical question that needs much investigation – I have in mind cases where a person's thought and behavioral processes are so disordered that he is incapable of understanding rules, weighing evidence, appreciating the consequences of his action, and controlling his behavior. In the absence of such impairment, mental illness ought not to be sufficient for involuntary commitment even with an accurate finding of dangerousness. Without such an impaired capacity to be guided by legal rules, dangerousness in the mentally ill ought to be dealt with as it is in the non-mentally ill – punished through the criminal process after illegal acts occur, but not used as a ground for loss of liberty before such acts occur.

II

If most current involuntary commitment processes are indefensible in important respects, as I believe they are, it is crucial that we correctly understand why this is so. I have already defended one important difference distinguishing civil commitment from penal commitment – the use of a preventive approach, preventive detention, with *some* mentally ill persons – and argued that differences in the civil and penal processes preclude any straightforward establishment of the proper protection of the non-dangerous in civil commitment from the bias toward protecting the innocent in the penal process. These are common lines of attack against involuntary commitment, but they are not decisive. Moreover, since our consideration thus far has been restricted to commitment on grounds of dangerousness, there is a whole class of involuntary commitment that we have not yet considered at all – those cases where the person is not even presumed to be dangerous. Thus, it may be helpful at this point to distinguish several different categories of potential involuntary patients (though it will not always be clear into which category any real person fits) and to summarize which if any of the earlier justifications offered for criminal punishment are relevant here as well.

1. Being Mentally Ill and Dangerous to Others

Here the "protection of others from harm" justification and, sometimes, the analogue of the rehabilitation justification, which I shall call effective treatment, are relevant. Commitment need not be unjustified in these cases in the absence of effective treatment, since the protection justification will remain, though where such treatment is available, an additional justification is present.

2. Being Mentally Ill and Dangerous to Oneself

Here the protection of others from harm is not relevant, though the protection of the patient from self-inflicted harm is relevant. I shall discuss in Section III the differences between these two principles – the harm to others and harm to self principles. Effective treatment, as with the previous category, may, but need not, be available as additional justification.

3. Being Mentally Ill where Effective Treatment is Available

Here, where the person is presumed non-dangerous, the only justification available will be the improvement that can be brought about in his condition by treatment, a provision of benefits justification.

4. Being Mentally Ill where Custodial Care is Needed and Effective Treatment is Unavailable

This, in fact, turns out not to be a separate category from two and three above, but usually some combination of them where custodial care will improve the quality of the person's life somewhat (a version of 3) and prevent minor harms (a version of 2). While those included under 2 above are generally suicidal and deeply self-destructive persons, in this category we are also frequently dealing with severe brain damage or senility where the potential harms are less dramatic and arise from the person's inability to provide for himself the ordinary necessities of daily life.

Several observations can be made. First, if harm to others is more serious than harm to oneself, then, assuming similar levels of predictability in each category, the strongest justification for involuntary commitment seems to exist for the first category – being mentally ill and dangerous to others. Second, the deterrence justification is absent in all cases since the action or behavior that is to be prevented with civil commitment is presumed not voluntary or responsible, and so will not be deterrable in the same way that criminal behavior will often be. If it is deterrable behavior, the person should be dealt with in the criminal, not civil, process, as discussed earlier. The dangerousness from which protection is being sought is presumed to be not in the mentally ill person's control in the sense in which criminal behavior is under the control of the non-mentally ill criminal offender. Deterrence is generally and rightly considered an important factor in the justification of criminal punishment, and its absence in civil commitment renders the latter's justification considerably more difficult. For example, suppose one year's imprisonment for the criminal act of reckless driving had the effect of deterring a significant number of other persons from driving recklessly who otherwise would have done so; then the total harms prevented will be several times that prevented by the civil commitment of one mentally ill person who will drive recklessly (even assuming perfect predictability) by reason of his mental illness, since in the latter case no one else is deterred from performing similar actions. This suggests in particular that because of the deterrent effect, acts causing relatively minor harms may justifiably be made criminal and carry prison terms, whereas prevention of the same act would not justify civil commitment of a mentally ill person for the same period. Third, the justice or fairness justification is always absent since civil commitment of the mentally ill is not for voluntary and responsible action. Finally, notice that the unqualified category of being mentally ill has been omitted above, though that is sufficient legal basis for involuntary commitment in some states. None of the justifications for penal commitment, or their analogues, are sufficient to justify commitment of persons simply because they are mentally ill; I will elaborate on this in Section V.

All the justifications of involuntary civil commitment that I have referred to above are of what I will call a utilitarian sort – they refer to the provision of benefits and the prevention of harms. If we accept a general utilitarian moral theory,[4] the justification for all treatment of mental illness, and for involuntary treatment in particular, will be determined by an overall weighing of the harms and benefits in such action to all concerned or affected. Commitment will be justified when, and only when, there is no other alternative course of action (with doing nothing as one alternative course of action) that produces a more favorable balance of harms prevented and benefits produced. I will shortly discuss an alternative, and in my view more acceptable, moral theory that would require even more restrictive conditions placed on the use of involuntary commitment. For the moment, assume that utilitarianism is correct; if we can show that involuntary commitment is rarely justified by it, then it will be even less frequently justified by the alternative theory. I have suggested above why the bias of the criminal law toward protecting the innocent need not be carried over to the same extent with the harmless in civil commitment, and why a preventive approach to harmful acts is justified with some, though by no means all, mentally ill persons. Can we prevent greater harms than we create by involuntary commitment of persons judged mentally ill and dangerous? If we can, then we have found plausible justification, and have cleared away some seemingly plausible objections, for doing so. The difficulty is that there is only rarely, if ever, sound reason to believe that we produce a favorable balance of harms over benefits with involuntary commitment. The problem lies not in the absence of a plausible moral basis for involuntary commitment, but in a failure to satisfy the empirical requirements of that basis.

The problem lies with the predictability of dangerousness. Dangerousness is rarely, if ever, predictable in the mentally ill with sufficient accuracy. Several points need to be made in this context. The first is that the category of dangerousness must be restricted and made precise. The complete loss of one's liberty involved in involuntary commitment is a very substantial harm. Acts to be prevented under the category of dangerousness must at least involve harms of the same dimension. Homicide, rape, and other serious acts of violence to persons are the appropriate sorts of acts to be prevented. Persons who exhibit bizarre behavior such as loud and incoherent talking to strangers in public places are frightening to many, and often a nuisance. Nevertheless, I believe it is clear that no harm that they cause comes close in magnitude to the harm of their loss of liberty if involuntarily committed. Offensive behavior, such as indecent exposure, also should be excluded from the category of dangerous behavior – being offended, while perhaps a harm, is again not a harm of sufficient magnitude; the same holds for behavior that is disruptive. The acts relevant under the category of dangerousness should be homicide and other acts causing serious physical or psychical injury.

The next point is a simple, but crucial, statistical one ([16, 21]). When the category is restricted in this way, dangerous acts are of quite infrequent occurrence. Even an extremely accurate test or predictive procedure for identifying people as dangerous or non-dangerous will produce far more false predictions that a person is dangerous (in statistical terms, false positives) than correct predictions that a person is dangerous (true positives). For example, suppose, not unreasonably, that only 1 out of every 1000 persons will commit a serious act of violence of the required sort. And suppose we have a test or predictive procedure that is accurate 99% of the time in sorting out persons as dangerous or non-dangerous (i.e., will have 1% false positives and false negatives). This is vastly more accurate than any predictive methods we now possess. Suppose 1 000 000 persons must be tested. This test will correctly identify 990 of the 1000 dangerous persons, and by involuntary commitment we could prevent the future harms they would cause if left at liberty. But the test will also identify 9990 other persons as dangerous who in fact are harmless. To prevent the harms that the 990 will cause, we must inflict on the other 9990 harmless persons the very serious harm of the loss of their liberty.[5] I do not see how any utilitarian calculus of benefits and harms could support involuntary commitment based on such procedures. Since our actual predictive techniques are considerably less accurate than those assumed in the example, the actual case for involuntary commitment on predictions of dangerousness is even weaker than the example suggests. Of course, if the mentally ill were far more dangerous than the public at large so that among them the incidence of violent acts was far greater, then the ratio of those in fact harmless to those in fact dangerous, who are identified as dangerous, would be reduced. However, available evidence suggests that the mentally ill are overall not significantly more dangerous than the general public [11, 19]. It may, however, be possible to identify narrower categories of mental illness, for example certain forms of paranoia, associated with a considerably higher frequency of dangerous acts of the required sort, though the considerable imprecision and vagueness in many categories of mental illness is a hindrance to establishing such associations. While sufficient predictive accuracy to justify involuntary commitment seems at present not to exist for many, if any, narrower categories of mental illness, it is only such narrowed categories that hold any significant future prospect for adequate prediction.

Since the harm of the loss of liberty to the person not dangerous is a very serious one indeed, the burden is on those who would use any predictive technique for dangerousness to establish that it possesses the necessary accuracy before it is employed in involuntary commitment proceedings. The point holds equally for any clinical method of establishing dangerousness – it must be carefully tested and the necessary accuracy established *before* it is used for commitment. I suspect that most clinical methods for predicting dangerousness have never even undergone adequate testing, much less established a sufficient degree of accuracy. Even for a utilitarian, it would seem that no test or predictive method producing more than one false prediction of dangerousness for each correct one would be admissible. The actual calculations required are more complex than I have suggested and must include such factors as whether and in what ways a dangerous mentally ill person can be made non-dangerous, what period of involuntary commitment is likely to be required, and so forth. However, I believe the conclusion stands that present or foreseeable predictive techniques for dangerousness are rarely, if ever, sufficiently accurate to justify their use to deny the mentally ill their liberty, certainly for anything beyond extremely short-term emergency hospitalization. Predictions of dangerousness could more easily justify involuntary treatment where it does not require confinement of the patient; this is simply an instance of the general principle that, other things equal, the least restrictive treatment alternative is always to be preferred.

III

The preceding argument assumed a utilitarian theory of moral justification, but even with it, involuntary commitment on grounds of dangerousness is very rarely justified. However, the general moral views of many people, and their views concerning civil commitment in particular, are not adequately accounted for by such utilitarian reasoning. In order to bring out the nature of these different views, and in particular to shed some light on the moral significance, if any, of the harm to self vs. harm to others distinction, I want to contrast with utilitarian moral theories, theories that have as basic moral principles the assertion of individual rights. The right relevant to the present question concerns a person's liberty, his right to make various choices for himself from available options, and to act on them without interference from others, choices about his life, about what form or pattern his life will take.[6] If persons have such a right, however, it is not unlimited and in particular it is limited in circumstances where one person's exercise of his right violates others' rights, or causes serious harm to others. But what distinguishes such a moral view as non-utilitarian is the fact that with it, rights cannot justifiably be infringed or violated simply because there is a bare gain in utility, in the balance of benefits over harms, in doing so. The strength or importance we give this right will be reflected in part by the loss of utility we are prepared to sacrifice in order not to violate it, but any non-utilitarian rights-based theory will require the sacrifice of *some* utility in order to protect basic rights. Utilitarian views of moral rights will make them derivative on their promotion of utility, and so justifiably overridden and infringed when they fail to do so; moral rights do no real work in a utilitarian theory. [I believe such a rights-based moral theory is ultimately more defensible, and better accounts for most people's considered moral judgments than does a utilitarian view, though this question is much too lengthy to take up here.] What I want to emphasize here is that if principles ascribing moral rights to persons are among our basic moral principles, we will not always be prepared to violate those rights, in particular a right to liberty, simply because doing so would produce a gain in utility, in the balance of benefits over harms. On such a moral view, the protection from harm to oneself or others, and the use of involuntary treatment to benefit its recipient, will justify commitment of persons judged mentally ill in a still narrower range of cases than on a utilitarian view, and will require still more accurate techniques for prediction of dangerousness.

Why one might hold such a rights-based moral view is related in part to the concept of autonomy. This concept is by no means a clear one, but it involves our view of persons as able to form purposes, to weigh alternatives according to how they fulfill those purposes, and to act on the result of this deliberative process.[7] This aspect of autonomy concerning our behavior being determined by this conscious choice process is something that many people value, and value not simply because being autonomous leads to maximized desire satisfaction or utility. Rather, it is part of an ideal of human excellence neither entirely subject to nor based on a utilitarian calculus. Moral rights, since they specify an area of behavior in which it is wrong for others to interfere with our acting as we choose to, at least without special and non-utilitarian justification, are particularly well-suited to protecting and expressing this value placed on individual autonomy and self-determination.

With this contrast between utilitarian and rights-based theories we are in a better position to understand, or at least to place in context, the dispute about whether harm to oneself is on a par with harm to others as a justification for infringing person's liberty in general, and for involuntary civil commitment in particular. John Stuart Mill laid great importance on the harm to self versus harm to others distinction when he stated that

> the sole end for which mankind are warranted, individually or collectively, in interfering with the liberty of action of any of their number, is self-protection . . . the only purpose for which power can be rightfully exercised over any member of a civilized community, against his will, is to prevent harm to others. His own good, either physical or moral, is not a sufficient warrant. He cannot rightfully be compelled to do or forbear because it will be better for him to do so, because it will make him happier, because in the opinion of others to do so would be wise or even right ([17], pp. 95–96).

On this view, though Mill later qualified it, *only* the prevention of harm to others justifies interference with a person's liberty, prevention of harm to the agent himself never does so. What kind of general moral theory or view justifies this very sharp moral distinction between harm to self and harm to others? Suppose the appeal is to a utilitarian view, whether utility is conceived in terms of pleasure and the absence of pain, human happiness, desire satisfaction, or whatever. Utilitarianism is what Ronald Dworkin has called a goal-based moral theory because it evaluates human actions, public policies and social institutions according to how they promote the production of the goals specified as valuable by the theory [6]. If utility is conceived in terms of human happiness, for example, it follows for a utilitarian that whenever we can best promote the human happiness of all affected over the long run by interfering with a person's liberty of action, we are morally justified in doing so. But if harms are generally productive of human unhappiness, then they will serve as at least prima facie grounds for interference with liberty, whether the harm is to the person whose liberty is interfered with, or to others. On a goal-based utilitarian moral theory, any basic moral principle justifying harm prevention will be a general one, making no distinction between harms to self and harms to others. Any difference in the two as a basis for justification of interference with a person's liberty of action must be based on a difference in their relative promotion of human happiness, and it is doubtful that they are substantially different in this respect. Mill's principle, as a basic moral principle, seems indefensible on utilitarian lines.

The moral relevance of the harm to self versus harm to others distinction (though not in as strong a form as Mill gave it) can, however, be defended by appeal to a moral view that takes the possession of individual moral rights, among them a right to liberty or privacy, as fundamental. Rights in general define areas of behavior in which their possessor is free to act as he chooses and as he sees fit without interference from others, even though his action may be regarded by others as unwise, distasteful, foolish, etc. Rights are not inviolable, though, and in particular may be justifiably infringed when doing so is necessary to prevent the violation of another person's more serious or more important right, that is, in cases of conflicts of rights. If rights are fundamental in the moral theory, then it is plausible as well to defend a somewhat different kind of right, sometimes called a defensive right – the right to be *free from* serious injury and harm caused to oneself by others. Prevention of serious harm to others could serve as justification for interfering with someone's liberty of action, and even his confinement or commitment if necessary in order to prevent a serious violation of that other person's right. Prevention of harm to oneself could never be justified as necessary to prevent violation of an analogous right of one's own – I can violate other's rights, but it is not possible for me to violate my own rights. If I have a right to x, then it is wrong, other things equal, for others to prevent my doing, or having, x if I choose to do so; it is not, on the other hand, wrong for me to fail to exercise my right to x, or to act in a way that will prevent my doing or having x. Preventing a person from harming himself can never be justified in order to protect his right not to be harmed, or injured, or killed; this is a right against others, and against the action of others, and so could never conflict with, and in turn override, one's right to liberty in cases of causing harm to oneself.

The second point relevant to a rights-based defense of the moral significance of the harm to self versus others distinction has already, in effect, been made. It is that rights mark off areas of behavior in which their possessor is to be free from the interference of others without some special justification. The promotion of general utility cannot be such a special justification for violating rights, or else there would be no point in insisting on individual moral rights, and their importance at all [5]. Rights specify areas in which it is wrong for others to interfere with our action even in some cases where overall utility would be promoted by such interference; they define areas of action in which what we do is 'our own business'. Thus, they represent constraints on the manner in which general utility or welfare may be promoted, either by individuals or the government. Acknowledging individual rights entails some sacrifice in the promotion of general utility, a sacrifice reflecting the independent value we place on human liberty, autonomy and self-determination. On a rights-based moral view, then, we are correct in at least some cases in not preventing a person from causing harm to himself where we would do so if the harm was to another; and the same, of course, holds for not committing a mental patient who is dangerous only to himself but not to others.

I should emphasize that the argument from the problem of predicting dangerousness is fully available in a rights-based moral view as well as a utilitarian one. In a rights-based moral view, it will make involuntary commitment on grounds of dangerousness to self *or* others more difficult still to justify morally. But assuming similar predictive accuracy of dangerousness to self or others, in a rights-based view, harm to self is a weaker ground for involuntary commitment.

IV

In the last three sections I have been principally concerned with a number of moral issues in the involuntary commitment of mentally ill persons on grounds of dangerousness. Much involuntary civil commitment, however, is carried out "for the patient's own good". This may involve the prevention of self-inflicted harm, the dangerousness to self category, but it need not. Categories such as "in need of care or treatment", "such that one will benefit from treatment", "in need of treatment and lacking sufficient capacity to recognize that need", are common in state commitment statutes and suggest justification solely on grounds of benefits to be provided to the patient committed. I want to explore the justification of conditions of this sort for involuntary commitment and treatment of mental patients. I shall suggest what such a paternalistic principle would be like on a moral theory that takes a right to liberty or privacy as one basic moral right. A utilitarian's moral principle of paternalism (at least, an act utilitarian's), will be relatively straightforward – acting towards some person for his own good will be justified, like other acts, if and only if doing so produces at least as much utility to all affected as does any alternative action open to the agent. The principal dispute for utilitarians will be, here as elsewhere, about just what are the effects on general utility or happiness.

We can use a recent definition of paternalism from Bernard Gert and Charles Culver [10], though with some relatively minor alterations.

A is acting paternalistically toward *S* if and only if *A*'s behavior indicates he believes that:

1. His action is for *S*'s good.
2. He is qualified to act on *S*'s behalf.
3. His action involves a willingness to violate one of *S*'s moral rights.
4. He is justified in acting on *S*'s behalf independent of *S*'s past or present consent.
5. *S* believes (perhaps falsely) that he (*S*) knows what is for his own good.

Assuming this is a largely adequate account of the nature of paternalistic behavior, of what such behavior is, when is paternalism morally justified? One way of posing this question is to ask when we would want others to so act on our behalf, when we would consider them justified in so doing. I suggested above that moral rights are intimately connected with the notion of autonomy or self-determination and the placing of important value on being autonomous. Our concept of autonomy as applied to persons depends on our conception of persons as able to form purposes, to weigh alternative actions according to how they fulfill these purposes, and to act on the result of this deliberative process. And being autonomous in this sense is something many people value for its own sake, as part of an ideal of human excellence, and not simply as instrumental to human happiness or some other goal. I believe our capacity for autonomy in this sense is a necessary condition for our being the sort of beings who could possess moral rights, and for it to make sense to ascribe moral rights to persons.

We will want others to act paternalistically towards us, (in ways that violate our moral rights), only when our capacities to form purposes, weigh alternatives according to how they fulfill those purposes, and to act on the result of that deliberation, are defective. Otherwise, we will want the protection of uncoerced choice and action that moral rights provide. As one philosopher has put it ([20], p. 249), we will want others to act paternalistically towards us only when we are subject to an evident weakness or defect of will or reason. In the absence of such failures of will or reason (a necessary though not a sufficient condition for paternalism), no belief that the person will be made better off by our action will be sufficient to justify our acting paternalistically towards him. Let me illustrate and say a bit more about what I mean by failures of reason or will.

By a failure of will I mean the inability to act as we believe we ought to act, that inability arising not from our being unable to perform the action successfully once we have undertaken to do so, but from a failure to be sufficiently motivated to undertake the action. Cases of extremely strong temptation, and physical or psychological addiction are examples – Odysseus anticipated his own weakness of will and lashed himself to the mast to prevent his acting on it. Cases of weakness of will provide perhaps the strongest basis for paternalistic intervention, but only because they border on, and in fact usually become, no longer cases of paternalism at all. If Odysseus was not untied because he gave prior instructions not to be, then this deprivation of his liberty is justified by his own prior consent to it; it does not require paternalistic reasoning. Failures of will that justify paternalistic intervention will have to be cases where we lack past or current consent for our paternalistic action, but nevertheless have reason to believe that the person is not able to act as *he* believes he ought to act, all things considered. We may act paternalistically when we have good reason to believe that the person's present action or wishes are based on a failure of will and that as a result he will in the future, either the immediate or more distant future, consent to our present action. We must have evidence that his long term or stable desires are in conformance with our paternalistic action, although his present desires and actions are not.

Notice that the reasonable expectation of future consent will not by itself justify paternalistic action, since much 'brainwashing' and coercive conditioning might satisfy this condition, but only the reasonable expectation of future consent *in addition to* independent evidence of the failure of will. Our reasoning in such cases should be based on our knowledge of the person in question and on an estimation of what he would consent to were it not for the specific failure of will affecting his present behavior. Showing that he will avoid some harm or receive some benefit will not be sufficient to establish that he would consent to our action, since it is exactly the freedom to make choices that are not the 'best' choices that rights are designed to protect. The harm prevented or the benefit to be received will have to be sufficient to make it reasonable to believe that this person, or, failing sufficient knowledge of him, any rational person, would wish others to act paternalistically towards him in this instance. On a rights-based moral theory, a principle of paternalism should in general require evidence that except for the defect in the person's reason or will, he would waive the right our action violates, that he would choose to have us act in a way that violates that right. I am inclined to believe that usually, where there is sufficient evidence of failure of will to justify the very substantial invasion of liberty that involuntary civil commitment represents, it will be possible to obtain the patient's consent to the hospitalization, certainly for anything beyond short-term emergency hospitalization. These will generally be cases where, as the patient subjectively experiences his situation, he feels unable to understand and/or control his behavior and does not identify with the behavior in the sense of wishing to be motivated in that way; the desire for help based on those feelings will generally be the principal evidence for what I have called a defect of will.

There are greater potentials for abuse in the application of the category of defects of reason, due to the practical difficulties in making some of the distinctions needed there. Consider first certain kinds of defects in knowledge. For example, a person is under the delusion that he cannot be harmed and is about to step in the path of a speeding car believing that he will not be harmed by doing so. Certainly, assuming we have no independent reason to believe he wants and has good reason to suffer harm, but only mistakenly believes that he will not be harmed, we would not hesitate to use coercion to push him out of the path of the car. We would be justified in this because the person is not getting what he wants in acting on his false beliefs. With a normal person, we could give him evidence on the basis of which he would then change his belief, and in turn give up the desire to perform the harmful action – no coercion would normally be necessary. Here, the defect of reason is the maintenance of false factual beliefs in the face of clear evidence of their falsity; coercion is needed so he can escape the substantial harm that he does not want and does not believe that he will receive.

The first difficulty, both theoretical and practical, in applying this category arises from the necessity of distinguishing failures of knowledge, defects in one's reasoning concerning factual or empirical matters, from differences in values. If what a person wants, and the kinds of activities he values, can only be different from those that others want or value, but cannot be 'right' or 'wrong', then it seems that the person with unusual values or desires has no deficiency of reason which could serve as justification for paternalistic action towards him. Some highly unusual desires are no more than that – highly unusual, perhaps even peculiar, but not irrational, for example, a taste for live toads or driving motorcycles across canyons. Some persons may simply enjoy eating things most find repulsive, or engaging in highly dangerous activities. Most differences in values and desires are of this sort, not subject to criticism as irrational, however unusual or even socially unacceptable they may be. This reflects the common sense view that there are differences in tastes and enjoyments that are perfectly natural, that make sense to us given what the person is like and what his plan of life is, and so do not justify criticism in terms of irrationality.

However, all desires or values are not simply different one from another. The objects of some desires are, in themselves, irrational. To suffer great pain, to be seriously injured, to be deprived of one's freedom or of significant opportunities, among other things, are intrinsically bad for a person, that is, they are intrinsic evils.[9] That is not to say that it may not be rational to desire one of these, for example if one's life can only be saved at the cost of serious injury. But to desire to be seriously injured for its own sake and for no further reason is irrational.[10] I would suggest that paternalistic action can be justified to prevent a person from acting on an irrational desire where the likely harm or evil, or loss of benefit, to him from doing so is very substantial.

Not only can the objects of some desires be irrational, but some weightings of gains and benefits, and of conflicting desires, can be as well. There is a subjective factor in the assessment of the amount or importance of a particular gain or loss that an activity will produce. Just consider the different relative importance different persons give to various features of products they purchase such as cars. People may reasonably differ in their reflections on such, so that there is room for reasonable difference in turn about the overall desirability of various choices, actions and activities. Such differences provide no justification for paternalistic action, for imposing one's own value-judgment on another. But weighing the relative importance of desirable versus undesirable aspects of an activity or action can provoke criticism from our notion of rationality. For example, a person who wants very much to live, but needs an injection to save his life, and refuses the injection because he is afraid of needles, is irrational because he is placing a temporary discomfort above his own life. The difficult distinction that must be made here, and which creates a very great potential for abuse in this category of paternalism, is between irrational and unusual desires and weightings of desires. Many persons (and many philosophers) will deny there is any distinction to be made here between rational and irrational desires, and weighings of conflicting desires or values, and so will reject any criticism of desires and values on grounds of their irrationality. I believe this disagreement about the concept of rationality is responsible for much of the dispute about when paternalism is justified. Certainly I have not provided an argument in enough depth to justify the concept of rationality necessary to my view. Nevertheless, I believe it is the case that the concept of rationality we possess does allow such criticism. Where paternalistic action is necessary to prevent action on such irrational desires that will result in serious harm or loss of substantial benefits, and perhaps assuming as well that there is good reason to believe that the person interfered with will later come to agree that his action would have been irrational, paternalistic action can be justified.

The account of the morally justified use of paternalism that I have sketched here makes use of some very difficult and problematic distinctions, and has nowhere near the precision desirable. I believe it is at least close to a correct account of the matter, but I want to add that given the very substantial potential for abuse in the employment of paternalism on the grounds I have suggested above, we might reasonably decide as a matter of public and legal policy not to grant anyone the right to act paternalistically, or to greatly restrict any such right, while nevertheless accepting the moral position I have proposed.

I believe the category of failures of reason, both in the form of maintenance of false factual beliefs in the face of evidence of their falsity and of possessing irrational desires, is extremely important in the area of mental illness. A considerable portion of psychosis is characterized by one, or probably more often some mixture of both these forms of failure of reason.[11] I would emphasize again that on the account of paternalism that I have proposed here the presence of these failures of reason is not itself sufficient to justify paternalistic action. Rather, such action is morally justified only when such failures of reason are present and they will cause their possessor to suffer substantial harm or the loss of substantial benefits in the absence of paternalistic intervention.

V

Finally, I want to discuss briefly what role, if any, the concept of mental illness ought to play in requirements for involuntary commitment. Under both emergency and non-emergency commitment laws in virtually all states it is necessary to establish that a person is mentally ill, whatever other conditions may be necessary, before he may be involuntarily committed. One form of challenge to the mental illness requirement, and to involuntary civil commitment generally, is the claim that mental illness is a myth, as Szasz has put it, and is not in fact illness at all. If mental illness is not illness, but is only a label we misleadingly put on something else, then it ought not to be a condition for involuntary commitment to hospitals, and any involuntary commitment arguably ought only to follow criminal offenses.

Now it is certainly the case that the concept of mental illness is much too vague, and there are wide variations in its application by professionals, much less the general public. But it can, I believe, be defended using the paradigm of physical illness, though when this is done much that is now often characterized as mental illness, by both professionals and non-professionals, would no longer be such. But even assuming that mental illness exists and is illness, 'being mentally ill' is not the correct requirement for involuntary commitment; if it is retained this should be because the commitment is in part for medical treatment, and a different condition that does the work that 'being mentally ill' now does must be added. The point is a simple one. However the concept of mental illness is plausibly defined and clarified, it is not the case that all persons who are mentally ill are at all times, or even most of the time, unable to make rational and responsible decisions about their own condition and the desirability of treatment. Yet it is presumably the assumption that being mentally ill makes it impossible to make rational decisions about treatment that accounts for our willingness to treat the mentally ill, but not the physically ill, involuntarily and for their own good. (Persons who are physically ill can refuse treatment necessary even to maintain their own life and others, or the state cannot generally intervene to force them to undergo treatment, though the account of the justified use of paternalism sketched above would apply to some cases of physical illness as well. And, of course, treatment decisions for physical illness are frequently irrational, on virtually any account of rationality.) It is a condition of incompetence, not mental illness that is needed – the discussion of paternalism should have made that clear. This condition of competence concerns a person's capacity to decide and act rationally and responsibly, that is, to weigh alternatives against each other, to consider the consequences of actions, information relevant to the decision, and so forth. The competence condition should rarely turn on the content of the decision about whether to undergo treatment voluntarily, since that will rarely be the only evidence available concerning the person's competence, and allowing it to serve as such has a great potential for abuse.

Competence (the capacity for rational decision-making), is clearly a matter of degree. Human beings are imperfectly rational at best – none of us is always rational in our choices and actions. Any condition of incompetence will be a threshold requirement – some level of incompetence should be required before paternalistic intervention is permissible.[12] If we keep in mind the vast range of irrational choices, including choices that pose serious dangers to their makers, that are not interfered with by the law, we will be wary of setting the threshold of competence too high. It should certainly be no higher than the *lower* end of the range of competence for rational decision found in non-mentally ill persons. Lack of competence then, and not mental illness, should be the necessary condition for all involuntary civil commitment, and if the requirement of a finding of mental illness is retained for other reasons, it should be supplemented with a condition of incompetence.

The question of whether mental illness should ever be a sufficient condition for involuntary commitment must now be reformulated as whether mental illness *and* incompetency together should be sufficient grounds for involuntary civil commitment. It is evident that they should not since both these conditions could be, and in fact are, satisfied in cases where the person's illness is not effectively treatable with current treatment knowledge and techniques. For a paternalistic justification of involuntary commitment, the person must be incompetent by reason of a defect of reason or will, mentally ill and such that there is a reasonable probability of successful treatment of his illness. This provides a quite straightforward justification for a rather limited right to treatment as well, though seemingly no less limited than that formulated by the Supreme Court in the case of *O'Connor vs. Donaldson*, 422 U.S. 563.

Where involuntary commitment is on paternalistic grounds of incompetence, mental illness and treatability, the involuntary hospitalization and provision of treatment is for the person's own good. If no treatment is provided, no benefit is provided to the patient, and so no justification of paternalistic intervention is available for the involuntary commitment. The Supreme Court decision in Donaldson supported this. It did not rule on whether there is any right to treatment where the person is committed on grounds of dangerousness. And in the case of dangerousness to others, protection of others from harm rather than any paternalistic principle may be the basis for commitment, though we have seen above just how weak a basis that is, using present predictive techniques.

I shall not attempt to summarize here all the various arguments of this essay. However, our examination of involuntary commitment of the mentally ill on grounds of dangerousness to others, and on paternalistic grounds, has yielded one rather unexpected general conclusion that is worth stressing. It has seemed to many commentators in this area that if involuntary commitment of the mentally ill is justified at all, it is most clearly so where the person is judged to be dangerous to others, whereas paternalistic justifications are commonly considered morally suspect. Paternalistic commitment is not devoid of important prediction problems of its own, but at least in cases of the provision of important benefits through treatment there may be a significant range of cases where those problems are considerably less severe than with commitment for dangerousness. Cases of paternalistic commitment may therefore be morally less problematic than cases of commitment for dangerousness to others.

Brown University
Providence, Rhode Island

NOTES

[1] The use of prison sentences for a range of years, such as 1 to 10 years, coupled with the present parole system, is an important qualification on this, though fixed-term sentences seem currently to be regaining favor.

[2] Persons often do spend significant periods of time in detention before trial, but except in limited circumstances, bail must be set for them, the payment of which will secure their release until trial.

[3] This view of justice is elaborated at great length and very powerfully by Rawls [20]; it is specifically applied to criminal punishment in, among other places, [18].

[4] Readers unfamiliar with the literature and issues concerning utilitarianism can consult any ethics textbook, or my paper [2].

[5] This example is taken with slight alteration from [15].

[6] In the law, much of what I include under the right to liberty is included under the constitutional right to privacy.

[7] For discussion of autonomy useful in the present context see [4].

[8] Dworkin uses the notions of goal-based and rights-based moral theories in [6].

[9] The examples of intrinsic evils are from [9].

[10] As I now see it, an account of the nature of irrational desires would appeal to the notion of persons as purposive beings, and to certain objects or states of affairs as useful to the promotion of the range of purposes characteristic of persons. (It should be noted that some contingent truths of human psychology will be needed here.) Very roughly, it will be irrational, other things equal, not to desire such things, or to desire their frustration. This, of course, is *far* too simple, but the present paper is not the place to take up this quite difficult question: I hope to say more about the nature of rational and irrational desires on another occasion. Among useful discussions by philosophers on this question are [20], chapter 7; [8, 1, and 9].

[11] More than one psychiatrist has maintained to me in discussion on this issue that desires to possess, for their own sake, things intrinsically bad for a person, what might be called pure irrational desires, do not exist. Their view is that seemingly pure irrational desires of this sort are always based on false beliefs (and on some views, in many cases unconscious beliefs) which if true would render the desire not irrational. Whether this is correct is, or at least should be, an empirical matter; even if correct, it does not show that irrational desires are not important in the area of mental illness, but only that pure irrational desires are not. In any case, I believe the category of pure irrational desires remains important for an understanding of the issues in this area.

[12] This point is made clearly in the very useful and comprehensive article [14].

BIBLIOGRAPHY

1. Brandt, R.: 1969–70, 'Rational Desires', *Proceedings and Addresses of the American Philosophical Association* **43**, 43–64.
2. Brock, D. W.: 1973, 'Recent Work in Utilitarianism', *American Philosophical Quarterly* **10**, 241–276.
3. Dershowitz, A.: 1968, 'Psychiatry in the Legal Process: A Knife That Cuts Both Ways', *Trial* **4**, 32.
4. Dworkin, G.: 1976, 'Autonomy and Behavior Control', *The Hastings Center Report* **6**, 23–28.
5. Dworkin, R.: 1970, 'Takings Rights Seriously', *New York Review of Books* **15**, 23–31.
6. Dworkin, R.: 1973, 'The Original Position', *The University of Chicago Law Review* **40**, 500–533.
7. Feinberg, J.: 1970, *Doing and Deserving*, Princeton University Press, Princeton.
8. Foot, P.: 1958–59, 'Moral Beliefs', *Proceedings of the Aristotelian Society* **59**, 83–104.
9. Gert, B.: 1970, *The Moral Rules*, Harper and Row, New York.
10. Gert, B. and Culver, C.: 1976, 'Paternalistic Behavior', *Philosophy and Public Affairs* **6**, 45–57.
11. Giovanni and Curel: 1972, 'Socially Disruptive Behavior of Ex-Mental Patients', *Archives of General Psychiatry* **17**, 397–398.
12. Hart, H. L. A.: 1961, *The Concept of Law*, Oxford University Press, Oxford.
13. Hart, H. L. A.: 1968, *Punishment and Responsibility*, Oxford University Press, Oxford.
14. Harvard Law Review Association: 1974, 'Developments in the Law – Civil Commitment of the Mentally Ill', *Harvard Law Review* **78**, 1193–1406.
15. Livermore, J. M., Malmquist, C. P., Meehl, P. E.: 1968, 'On the Justifications for Civil Commitment', *University of Pennsylvania Law Review* **117**, 75–96.
16. Meehl, P. E. and Rosen, A.: 1955, 'Antecendent Probability and the Efficiency of Psychometric Signs, Patterns, and Cutting Scores', *Psychological Bulletin* **52**, 194–216.
17. Mill, J. S.: 1951, *Utilitarianism On Liberty – Representative Government*, E. P. Dutton and Company, New York.
18. Murphy, J.: 1971, 'Three Mistakes About Retributivism', *Analysis* **31**, 166–169.
19. Rappeport and Lassen: 1965, 'Dangerousness – Arrest Rate Comparisons of Discharged Patients and the General Population', *American Journal of Psychiatry* **121**, 776.
20. Rawls, J.: 1971, *A Theory of Justice*, Harvard University Press, Cambridge, Massachusetts.
21. Rosen, A.: 1954, 'Detection of Suicidal Patients: An Example of Some Limitations in the Prediction of Infrequent Events', *Journal of Consulting Psychology* **18**, 397–403.

SECTION IV

THOMAS SZASZ'S PROPOSALS: A RECONSTRUCTION AND DEFENSE

CORINNA DELKESKAMP

CRITICAL USE OF UTILITARIAN ARGUMENTS: SZASZ ON PATERNALISM*

Thomas Szasz's criticism of present public policies with regard to what he calls behavioral deviants must be reconstructed on the basis of utilitarian reasoning, if the criticism is to make sense. If this thesis can be defended, then a direction for future discussion of problems presented by those policies is indicated. After all, the authors contributing to that discussion within the present volume have either ignored or failed to consider adequately the radical, philanthropic, and influential theories of Szasz that happen to oppose their reasonings.

Indeed, one is reluctant to blame people for shunning controversy with Szasz. The manner in which that author presents his ideas is apt to discourage anyone endeavoring to take him seriously.[1] Yet his ideas deserve a more respectable presentation, which uncovers the political significance of his position and deepens our philosophical understanding of the problems involved. The wish to provide such a presentation has motivated my attempt to undertake a utilitarian reconstruction of Szasz's arguments,[2] – a reconstruction, however, that will require the development of an understanding of utilitarian criticism able to the task.

My reconstruction will consist of two parts. Part I discusses the three most important public policy issues raised by the authors of this volume and the attendant criticism of Szasz. Part II attempts to defend the utilitarian nature of that criticism.

At the same time these two parts will correspond to two aspects under which the subject of the present volume and the relevance of Szasz's position must be examined. On the one hand, the 'public policy' of the title refers to present United States policies in dealing with the mentally ill. Thus the essays here collected have reached either critical or justificatory conclusions with respect to those particular policies. On the other hand, these essays present philosophical perspectives concerning those policies. So their particular conclusions must have been reached on the basis of some general understanding of the rules by which such policies ought to be evaluated. Concerning the first aspect, it will be seen that Szasz's evaluative conclusions appear neither coherent nor unique. Concerning the second aspect it can be shown that their seeming incoherence disappears and that Szasz reasoning indeed presents a significant alternative.

B. A. Brody and H. Tristram Engelhardt, Jr. (eds.), Mental Illness: Law and Public Policy. 177–207.

I

Three main issues concerning mental health and public poicy are discussed by the contributors to this volume: (1) the medical interpretation of behavioral deviance; (2) the grounds for the policy of non-voluntary commitment on the basis of mental illness; and (3) the justification of the 'insanity defense' in criminal justice. In examining each issue, I shall (a) determine its relevance with respect to the theme of this volume; (b) characterize the main positions of the contributors and opposing arguments by Szasz; (c) expose the *prima facie* utilitarian nature of Szasz's arguments; and (d) point to a difficulty that hinders making *'prima facie* sense' of that position.

1. The Medical Versus the Social Significance of 'Behavioral Deviance'

(a) If behavioral deviance is essentially a medical problem rather than a personal, social, economic, or political one, then the solution to the problem lies in the field of health care policy. This field, then, provides the setting in which the public policy debate takes place.

(b) Of the six authors entering that debate, only Margolis explicitly argues for a common presupposition: that the classification of behavioral deviance as a medical problem is to be preferred over any other account. He defends his thesis by claiming that Szasz's opposition to the medical model is based on a misunderstanding of the nature of medical knowledge.

However, it can be shown that (1) Margolis misunderstands the depth of Szasz's position, and that (2) his own 'ideological' interpretation of medicine fails to do the *ad hominem* job for which it is designed.

(1) Margolis's misunderstanding of Szasz's meaning becomes clear from his restatement of that meaning.

(i) According to Margolis, Szasz finds something 'inherently invalid' about the notion of mental illness, or – paraphrasing – 'something not conceptually coherent' or 'intrinsically incoherent' [13]. But what Szasz finds inacceptable has nothing to do with internal contradictions, but rather with the fact that 'mental illness' suggests an incomplete understanding of the significance of phenomena connected with behavioral deviance. This notion therefore veils rather than reveals what such deviance is about. Medicine understands these phenomena in view of disease patterns they are taken to exhibit. These patterns are considered as (conceptual) objects sufficiently determined in themselves ([24], p. 45). Non-medical psychiatry, on the contrary, seeks to understand these phenomena in view of problems lying behind them. These problems have (in turn) a social significance, such that it makes sense to speak of the validity of those problems in view of that significance. Thus, a dimension of meaning is added which cannot be captured by the medical view.[3] This meaning concerns objective states of injustice, rendering those oppressed by them helpless to such a degree that they dare not express their discontent openly ([24], p. 119). This is why they use instead the indirect or metaphorical language of physical complaints in order to ask for spiritual help. Thus, Margolis's stress on incapacitation as a result of mental as well as physical illness is not helpful to the issue at stake. According to Szasz, psychiatric incapacitation is distinguished by a semiotic function ([24], p. 10), which Margolis simply misconstrues when he paraphrases 'mental patients are always shamming' ([13], p. 3).

(ii) According to Margolis, Szasz finds the thesis that mental problems respond to medical treatment socially pernicious [13]. But the example Margolis provides for what 'socially pernicious' could mean, – namely 'therapeutic ineffectiveness' – indicates that he has much too harmless a notion of that predicate.

Szasz's choice of a non-medical interpretation of psychic difficulties is motivated by the risk of political suppression he has exposed in the existing medical systen. Since at least some behavioral deviances result from moral and social conflicts reflecting larger societal problems, a purely medical view reduces even the effects of political grievances to merely personal sickness ([24], pp. 8 and 25). Moreover, the institutionalization of medical care provides society with a political means of social control by silencing criticism ([24], pp. 67f, 213ff, 260; [25], pp. 5ff, 16).[4] The universal prejudice according to which science is a realm of 'objective', unquestionable facts helps to conceal the abuse of power manifested in such suppression. It discourages doubt in the moral justifiability of what is hypocritically termed public mental health care.

Thus Margolis's ideological interpretation of the nature of medical knowledge in general is not helpful to the issue raised by Szasz, as long as this interpretation is not publicly endorsed. Contrary to Margolis's opinion, it is not Szasz's own misunderstanding of medicine as a pure science which is at the bottom of his attacks, but his awareness of such a misunderstanding on the part of the public ([24], p. 26).[5]

(2) Margolis holds (arguing against Boorse and others) that value implications are operative not only within non-medical psychiatry (as Szasz assumes), but equally within medicine in general. According to Szasz it is the

common ignorance of this very fact that renders medicine a ready tool for given societies to abuse their institutional power for the suppression of potential criticism and for hiding the injustice involved in such suppression. Consequently, one might imagine that Margolis's effort in unveiling medicine's susceptibility to political influence could contribute to a more proper public understanding of the need for safeguards.

Yet it is just that desirable consequence that cannot be made to follow from Margolis's premises. Values are for him 'ideological' if they contribute to the self-understanding of a particular social system. In this sense the only disadvantage attending 'ideologically biased' medicine consists in the fact that a behavioral problem would be declared a mental illness in one society, but not in another. Szasz, however, is concerned not merely with contingent cross-cultural differences nor with inconveniences of such a harmless nature. Rather, he claims that within the framework of a given culture injustice is done by social institutions to those who have a just social cause for despair. The truly obnoxious aspect of medicine's value-implications lies in the fact that the care provided in such cases will seek to re-adjust the patient to that very set of social norms whose validity he had come to deny. By maintaining the language of "care for those in need", Margolis indicates that he considers 'social maladjustment' to be exclusively a problem of the maladjusted, whereas Szasz considers it as a symptom for some societal ill as well.

(c) The *prima facie* utilitarian nature of Szasz's argument derives from the fact that, confronted with two possible interpretations of behavioral problems (one medical, one social), he chooses the one whose endorsement will serve to unveil, and thereby render ineffective, what he takes to be a method of social oppression. Thus his very conceptual choice constitutes a course of action that is taken in view of promoting the freedom of citizens to choose their own causes of suffering from social evil and thus their own terms of happiness.[6]

(d) Yet within the social interpretation of behavioral problems chosen by Szasz an ambiguity can be discovered. Socially speaking, those problems are presented as 'moral' problems in two very different senses of the word. On the one hand, they signify a state of social injustice – and thus an immoral society – that victimizes people to such a degree as to incapacitate them in their human and social relationships. On the other hand, 'mental illness' presents a route of escape that – even though unconsciously ([24], pp. 112f) – is chosen by those who have despaired of reaching their goals in an open-handed way. Thus it is they who immorally use the advantages ([24], pp. 25, 133. 213ff, [25], p. 23) presented by a socially accepted sick role in order to realize their unfair intentions. While 'mental illness' is on the one hand an indirect mode of communicating social problems, it is on the other hand a result of some choice in favor of communicating indirectly. Behavioral deviance, then, is not only a sign for a particular social ill, but equally the result of an 'action' by which the agent seeks to procure what he believes to be his advantage ([24], p. 56).

As a result of this double interpretation, it remains unclear what could, on utilitarian grounds, justify accepting either side, and it remains uncertain who, in the end, is to blame.

2. The Contingent-Property Versus the Necessary-Assumption Account of Personhood

(a) Social and legal systems function on the assumption that its members are persons, i.e., among other things, rational beings capable of pursuing what can conceivably constitute their happiness and of weighing the costs and benefits of any course of action taken. According to the medical model for interpreting deviant behavior, there exist human beings who are not in that sense capable, and hence not persons whose free decisions one ought to respect. With regard to such non-persons special social and legal policies are then required. While those legal policies concern the criminal (insanity defense) as well as the civil law (involuntary commitment and treatment), I shall restrict my present attention to the latter, reserving the former for the third issue to be discussed later.

(b) Most of the authors[7] presented here have defended or have at least implicitly endorsed a number of criteria by which to single out those cases of mental illness that would require involuntary commitment (for paternalistic or social reasons). The very admission of such criteria is criticized by Szasz on two counts. First, such criteria are based on a concept of rationality.[8] Rationality here denotes the capability of acting on behalf of one's own interest, where that interest has a relation to one's respective happiness. Yet any list of acceptable ideas of and ways in pursuing happiness reflects the particular preferences of a given social system.[9] In addition, such criteria will serve to distinguish those who count as persons (and are granted the freedom of acting according to their own choices) from those who do not. As a result, once the validity of any such particular list concerning rational behavior is accepted as a general account of how persons behave, it will prove an effective instrument in silencing those whose sufferings deny that validity by declaring them incompetent to criticize.

Second, the very act of psychiatric classification entails a prediction concerning human behavior. Yet an essential part of man's understanding his own identity – and hence his view of his options in acting – depends on the way in which he is regarded by others. Such a prediction, while supposed to be merely the outcome of an act of cognition, in effect functions as a new determinant of that behavior. People classified as insane are thereby impelled to conform to the expectation suggested by that role; if they resist, their case will be considered 'still more serious' ([25] pp. 190ff, 202ff, 216).

(c) The *prima facie* utilitarian nature of Szasz's argument derives from the fact that, confronted with two possible interpretations, he chooses the one that will provide an efficient safeguard against the abuse of paternalistic reasoning. The two interpretations concern the act of classifying someone as insane (as a non-person), one of them cognitive (modelled on the way in which the biologist classifies an animal), and one effective (modelled on the way in which people influence the actions of others on the basis of acting according to their opinions concerning those actions). By choosing the latter, Szasz can infer that the act of so classifying is unjustifiable in itself, and hence he can ensure that no particular criteria of classification should be used for social or political suppression. 'Personhood', then, denotes a necessary assumption one is to accept in every case.

(d) Yet within the interpretation chosen by Szasz a new ambiguity is discovered. On the one hand, the act of classifying someone as person or non-person determines the way in which that one classifies himself, and thereby determines his capability or incapability to behave as a person. Thus, insofar as persons are social beings, their personhood is brought about by a society treating them as persons. On the other hand, Szasz's moral disapproval of people who escape into mental illness rests on the assumption that such "illness" is the product of some goal-directed behavior in the first place. Such behavior – so Szasz assumes – is directed towards procuring happiness for the agent (even though in a somehow unwise manner). It differs from other kinds of goal-directed behavior only insofar as it is unconsciously goal-directed and as the method chosen is indirect. As a consequence, 'rationality' or 'personhood' is not the product of social interaction, but is a given fact about behavioral deviants.

While from the interpretation initially described there exists a moral obligation to treat someone as a person in order to let him be one, on the second account there exists a factual obligation to treat the mentally ill as persons, because, indeed, they are. As a result, it remains unclear what sort of property personhood actually is.

3. Moral Charity Versus Utilitarian Benevolence as Evaluative Grounds for the Insanity Defense

(a) If behavioral deviance is a sign of mental illness, and if some forms of that illness take away freedom, then as a matter of public policy concerning the criminal law the conditions must be determined that allow for the so-called insanity defense. In addition, the state's obligation to protect citizens from being harmed by others necessitates further policies concerning the subsequent confinement and/or treatment of the 'dangerous' ones thus excused from liability.

(b) Szasz's opposition to the depersonalization effected by psychiatric classification is met with the claim that only on the basis of such classification can the insane be excused from delinquency, a policy agreeing with our most basic moral intuitions. "A person is not to blame for what he has done if he could not help doing it" ([20], p. 75), Neu writes, and Moore, Brock and Coleman agree.

There are three reasons that can support Szasz's counter-claim that it is, all things considered, still more justifiable not to excuse the 'insane'.

(1) Institutions for the criminally insane are generally admitted to be less pleasant than prisons. Moreover, the period of detention often exceeds the duration of a criminal sentence. Hence, even for the one excused on the grounds of insanity, penal commitment would be the more charitable choice.

(2) Criminal punishment is often seen to imply society's moral disapproval of the delinquent. The insanity defense is said to spare him such reproach. Yet this advantage may be challenged both by a regard to practice and to principle.

(i) Practically speaking, it is hard to say whether the reformed criminal or the madman discharged as healed will meet with greater prejudice and suspicion when trying to return to normal life.

(ii) With regard to the issue of principle, the authors of all relevant essays agree that being punished is painful and that the benefit (whether deterrence or prevention) lies only on the side of public interest. They neglect a second moral implication of punishment as a chance for the delinquent to 'atone' for his violation of the public order. His responsibility being admitted, both the perpetration and the atonement are credited to him. None of this holds, on the other hand, for the criminally insane. He is a loser both in 'pleasure' and dignity,[10] – a result that squares unevenly with the charitable motivation behind such treatment.

(3) Denying that behavioral deviants act responsibly and referring their

problems to the medical realm render ineffective the critical significance such behavior could have for society ([24], p. 24). Following this line of thought, one would have to conclude that the practice of punishing even those deemed insane, instead of obscuring the social conditions from which their plight ensued, would openly expose the injustice of the existing social system. Thus, the punishment of psychic deviants, just as now the punishment of the socially disadvantaged, eventually would create sufficient scandal to motivate reform.

(c) Szasz's three-fold argument exhibits two different kinds of *prima facie* utilitarian reasoning.

In (1) and (2) on the basis of a more complete assessment of the implications and consequences of the insanity defense, Szasz argues for the opposite policy, of not exempting from liability, adducing among other reasons that the fate of those suffering from mental illness could thus be improved. Yet this reasoning rests on two presuppositions Szasz himself would not endorse: (i) that one ought to take a charitable course with insane criminals because they are really ill, (ii) that the two policies open to choice: excuse-and-commitment-because-of-danger-to-others and holding-responsible, give an exhaustive account of the relevant consequences. Hence, the first aspect of that reasoning seems for the moment to make sense only in the context of an ironical *ad hominem* argument and can therefore be neglected.

In (3) Szasz can be taken to respond to a problem that might be found hampering his own recommendation: How would that policy agree with the moral implications of applying the law to those who have not acted freely? Is it, in other words, right to punish them? Szasz's response would then follow the utilitarian maxim that questions concerning rights and morals must be decided in view of the advantages incurred.

(d) Yet that line of reasoning, if it is to be conceived in utilitarian terms, presents a further difficulty. In what sense of 'utilitarian reasoning' should it be plausible to recommend a policy on the grounds of the scandal it will present? Given that scandals and public outrages are complex social phenomena, it still seems legitimate to assert that they are socially harmful, and thus anti-utilitarian in their effects.[11]

Looking back at the three issues discussed, it appears that the central problem underlying them concerned the justifiability of paternalistic action by the state with reference to behavioral deviants. Thus the second issue dealt with the conditions under which someone can become subject to such an action: the first and third issues considered the actual 'benefits' received. While the authors have essentially restricted themselves to devising various kinds of justifications for various kinds of paternalistic interference, Szasz has opposed that very attempt at justification ([24], pp. 176, 258). Proceeding on the principle that every one should be granted freedom to pursue happiness in his own terms, he claimed that paternalism of the kind discussed is not a means to secure someone's happiness. It rather presents a hindrance to his pursuing it by himself, in any unwise or deviant manner of his choosing, and thus a hindrance to his experiencing consequences that may teach him to become more efficient.

To consider such individually determined happiness (of the greatest number) as the ultimate value is clearly a mark of utilitarian thinking. It is this consideration that will be seen to have motivated not only Jeremy Bentham's criticism of any moral grounds for the validity of legal systems, but also his endeavor to propagate instead their justification on the basis of resulting happiness. It is that same consideration that motivated John Stuart Mill's critique of public opinion as restricting what counts as acceptable ways of seeking happiness ([15], pp. 9f, 77). Hence it appears that our project of rendering Szasz's arguments respectable in the eyes of the philosophical community by reconstructing them in somehow utilitarian fashion was particularly fitting not only with respect to his 'political' view of the subject of his arguments but also with respect to their critical conclusions.

However, as the need for that reconstruction arose in the context of a discussion concerning evaluations of present paternalistic policies, the project appears nevertheless to have been pointless. Brock, reasoning not from a general critique of paternalistic action (after all, he believes in mental illness), but from a lack of criteria for distinguishing those to whom such action could justifiably be applied, presents rather the same negative conclusions concerning those present policies as does the entire Szaszian theory. Moreover, our examination of that latter theory has revealed its internal inconsistencies. Finally, Brock's essay furnishes arguments suggesting that any theory of justice based on the notions of desert and individual rights, and thus the theory presented by Szasz, must be a non-utilitarian theory. As a consequence, even if it could be proven that the authors of this volume missed the point of Szasz's arguments, there was nothing much to miss, since these arguments are neither respectably utilitarian, nor consistent, nor do they lead to any new conclusions.

II

The first part of this essay focused on the manner in which present paternalistic policies have been evaluated. The second part will now examine some underlying theoretical frameworks suggesting the general principles by which such evaluations ought to be conducted. I shall no longer consider ways of judging concerning present policies, but ways of justifying the manner of such judging.

The main design of this essay is to make sense of Szasz's arguments by exhibiting their underlying utilitarian reasoning, and there is now only the second, philosophical level of the discussion left to achieve that goal by means of an understanding of 'utilitarian' that is devised to permit such an interpretation. Thus my examination of theoretical frameworks legitimizing possible evaluative conclusions, and, in particular, of the utilitarian framework legitimizing Szasz's conclusions, will take issue with only one of the authors presented in this volume. In response to Brock's results, contesting not only the relevance but even the utilitarian nature of Szasz's argument on the practical level of the discussion, I shall (1) show that on the philosophical level Brock's theoretical framework fails in a manner in which Szasz's account does not, (2) outline an understanding of utilitarian criticism that, contrary to the one indicated by Brock, will reestablish the utilitarian character of the arguments construed for Szasz, (3) use the results of that understanding as an interpretive tool by which to resolve the inconsistencies attending Szasz's position and to establish its political relevance.

1. The Adequacy of the Theoretical Frameworks Underlying Brock's and Szasz's Conclusions

A theory regarding the state and its legal system is adequate for the task of suggesting principles by which to evaluate paternalistic policies if it provides safeguards against the abuse of such policies. In particular, those accounts that allow for the possibility of justified paternalism must furnish reliable criteria for isolating the cases where that possibility is realized. Reliable criteria of the sort required here are such that no particular social or political situation can be imagined in which those criteria fail to work.

The criteria Brock has given are designed to single out those human beings who are incapable of acting in their own interest [8]. In order to show that these criteria are no reliable guides to justified paternalism, we do not even have to strain our imagination. Even a superficial analysis of the Spanish Inquisition's policy of torturing supposed witches and of the quasi-paternalistic justifications given for that policy provides a historical precedent where Brock's criteria are seen to tolerate an indefensible practice.[12] By pointing to that somewhat sinister example I am not suggesting that another

17th-century witch hunt might occur again in the future. I am only stating that no criteria can be relied upon in principle if they would have failed in a particular instance in the past.

Thus it appears that at least one reason leading to the unfavorable conclusion of the first part of my essay does not hold true for the second. Philosophically speaking, the similarity of practical conclusions drawn by Brock and Szasz becomes irrelevant if the reasons given by Brock are not satisfactory.

However, it might be thought that the theoretical framework grounding Szasz's conclusions is also unsatisfactory in the same regard. Let us, therefore, examine what such a failure could possibly look like.

As Szasz's theory does not even admit the possibility of justified paternalism, there are no criteria that could be proven unreliable. The only remaining attack consists in challenging his very anti-paternalism as a sophisticated sort of paternalism, and in revealing how it might come to hinder someone's pursuing freedom in its own quasi-deviant way. One might argue that Szasz's moral imperative to treat all people as persons, i.e. to respect their freedom and to consider them responsible for their actions, will induce a state to disrespect the freedom of those who have chosen to pursue the happiness of mental illness, irresponsibility and unfreedom within the confines of an insane asylum and who, while enjoying their nonliability, fail to care about the personhood denied them.

But such an argument is invalid. Disrespecting someone's free decision to run after a rather odd kind of happiness by calling him irrational is not at all like disrespecting someone's free decision to want to be considered unfree. To an unprejudiced mind there is nothing inherently inconsistent about freely choosing strange delights, but there is something illogical about freely choosing one's being constrained, for example. Of course there is nothing illogical, on the other hand, in freely willing to be committed to an asylum where one will subsequently be held captive, or even to be so committed and to appear constrained in the course of that commitment. What is illogical is only that kind of choice where the form is explicitly denied by its content. To be sure, no state can be accused of unjustified paternalism if it does not respect someone's desire to merely appear unfree while really remaining free, since no state is obliged to respect a lie. Similarly, no state can be so accused if it holds people, who may now be unfree, responsible for the act by which they freely forsaked their freedom and thus for all the consequences (with certain qualifications) they might incur. It is in order to assimilate what otherwise might appear as a will willing its own not-willing (an illogical project that must therefore be impossible to realize) to the more harmless patterns of lying or acting irresponsibly, that Szasz undertakes to decipher deviant behavior in terms of unconscious goals and advantages indirectly pursued. A state, while acting on the basis of such deciphering, cannot be said to constrain anyone's liberty, for the simple reason that a liberty to do the impossible does not exist. As a result, Szasz's justificatory framework for evaluating present policies is superior to the one given by Brock in such a way that a renewed effort for making sense of Szasz's position appears worthwhile after all.

2. Utilitarianism as a Critical Endeavor

If Brock is right in excluding the notions of 'moral desert' and of 'individual rights' from utilitarian theories of justice and of the state, then my reconstruction of Szasz's arguments has failed to render them utilitarian. By showing in what sense Brock can be taken to be wrong I shall try to defend my main thesis asserting the necessity, and thus the possibility, of such a reconstruction.

There are among many others two thinkers commonly associated with utilitarianism, Jeremy Bentham and John Stuart Mill. Brock concedes that Mill's theory has indeed a place for moral desert and individual rights – what he, however, denies [8] is that Mill's theory is utilitarian in a manner comparable to the manner in which Bentham's is. In response to Brock, I shall argue that (a) if Bentham's theory (1) has no use for moral desert as implied in legal punishment, and if it (2) does not place the same emphasis on individual rights as does Mill's theory, then this is not because these notions are incompatible with his utilitarian endeavor as such. I shall also argue that (b) Mill's theory is indeed as utilitarian as the one of Bentham and that it appears so different, both because of its different historical context and because of the fact that Mill – unlike Bentham – does not restrict his task to giving merely a negative response.

(a) (1) Bentham indeed opposes the thesis that legal punishment is somehow 'deserved' ([5], p. 23). Yet his criticism is not – as Brock would have it – motivated in principle because moral considerations have no place in a utilitarian theory taken as a whole. His opposition instead functions within the particular utilitarian argument that is directed against what he considers prevalent political abuse of such moral terms for the sake of an apologetic dogmatization of the existing body of laws. He finds these terms employed in such a way as to preclude critical scrutiny of the extent to which the state should be thought entitled for establishing legal sanctions for the regulation of conduct, and in such a way as to provide a semblance of justification for policies that in truth serve only the interests of particular factions.[13]

But Bentham's devising such anti-moralistic criticism does not entail that, in the *theoretical* framework underlying that criticism, moral considerations are lacking. Any such framework contains two parts: one defining the standpoint from which that criticism is announced (or the *position* of the author) and the other specifying the meaning of concepts that are used in the formulation both of the particular critical argument and of the author's position (or a *theory* on the kind of world in which the utilitarian criticism takes place). Thus Bentham's utilitarian *position* can be considered to be itself a moral position[14] (in the sense of a position taken within moral philosophy or, as I shall also call it, a meta-moral position). After all, by assuming this position he declares the promotion of the greatest happiness of the greatest number to be the *summum bonum* that should guide any particular moral consideration. Accordingly, Bentham argues for this position in a (meta-) moral way by claiming that to infringe upon someone's liberty, and thus (on the basis of Bentham's anthropological commitments spelled out by his theory proper) by implication to hinder someone's pursuing his own happiness, is the same as to do him wrong.

In view of the particular historical situation he confronts, this (theoretical) position entails a backing for his particular utilitarian argument noted above: in whichever social system 'morality' is placed in a realm 'apart' from considerations of resulting happiness, the fundamental moral principles, as well as the need to enforce them, must be presumed to be self-evident (or else ordered by some deity whose authority is self-evident). What is self-evident is not open to critical discussion. Hence those principles become easy instruments for concealing party interest behind the appearance of concern for a general good, which the governing class presumes not only to determine for but also to impose on the common people ([2], pp. 8f). The principle of (general) utility and hence Bentham's criticism of an apologetic moral theory of punishment is designed to draw such political stratagems into the light of rational examination and to expose the merely partial utility-considerations (mere expediency) behind them.[15]

This critical motivation, moreover, would not by itself necessarily exclude the possibility – as far as Bentham's own theory regarding the state and its legal system is concerned – of attributing moral desert to penal measures, as long as criminal acts might be taken to oppose the greatest happiness of the greatest number. The reason why such an account is not included lies in the fact that Bentham restricts the function of his theory to the grounding of the very particular critical argument with which he is concerned. As a result, Bentham's account of human nature is not necessarily meant to be complete but to provide society with an unfailing safeguard against ideological impositions from above. Hence he adduces only what he takes to be the most basic and therefore indubitable facts concerning the conditions of human happiness. These conditions are defined in terms of pleasurable sensations that, while encompassing both the egoistic and the sympathetic concerns, leave no room for purely spiritual values as envisaged by dogmatic morality.[16] In particular, Bentham is not interested in having his theory of human nature and happiness include an account of his own utilitarian endeavor as of a human endeavor itself that must, after all, be somehow related to human goals. Nor does it seem he feels that he ought to account for the moral connotations of that very undertaking. This is why his legal theory can do without considering specifically moral norms altogether, and hence without including any notion of desert in punishment.

Hence it is not because of the utilitarian nature of Bentham's undertaking, that moral desert is disregarded, but because of the historical situation from which his critical intention arises and from the fact that Bentham restricts his task to an adequate response to just that situation.

Mill, on the other hand, takes pains to argue for a moral desert theory of legal punishment in order to respond to those critics of the utilitarian position who claim that that position is incompatible with an adequate theory of justice. It is this criticism, directed more specifically against the theory underlying Bentham's particular argument, which, among other reasons, prompts Mill to devise not a minimal but a complete account of human nature, happiness, the state, and so on. Within this account he admits the distinction between deserving and not deserving punishment to lie "at the bottom of the notions of right and wrong" ([16], p. 454) and thus at the bottom of the legal system.

The desert Bentham opposes belongs to a system of morality as pretense for political interests. The 'desert' Mill endorses arises from a moralized version of the natural instinct of vengeance ([16], pp. 249, 259), where the notion of 'moralized' agrees with the utilitarian (meta-) morality of Bentham's position. The manner, however, in which Mill conceives that 'moralized instinct', reflects a second and purely philosophical reason for his endeavor to provide a complete theory rather than an account that merely serves a particular critical intention. Contrary to Bentham, Mill does indeed account for his own utilitarian position with its specifically (meta-level) moral connotations within his (ground-level) theory. Thus, the givens of human nature are not merely listed as such, but are placed in a relation to their possible moral perfection. Accordingly, the instinct of vengeance is acknowledged as a fact about how humans feel, but that instinct – insofar as it is rendered subservient to the principle of utility – is also acknowledged to be the foundation of the notion of justice which humans ought to entertain. In this way Mill supplements Bentham's criticism of a false theory of deserts by outlining a true one, which now, however, encompasses both private morality and the law. The moral aspect of legal punishment, which for Bentham has only polemically served within his utilitarian criticism, is by Mill included into the legal theory (in his underlying framework) as well.

Within his theory, Bentham holds that the state is morally and therefore legally obliged to promote the happiness of its citizens not 'conjunctively', as if to benefit an imaginary 'whole' over and above the sum of all the individuals' happiness, but 'disjunctively', for each individual ([5], pp. 184f). To be sure, it is supposed that each person knows best how to seek those pleasures that define his happiness. The task of 'promoting happiness' is thereby restricted to providing boundary conditions that allow for such individual seeking to take place. The right of each individual to pursue happiness is thus implicitly acknowledged. Contrary to Brock, a utilitarianism of such 'disjunctive' character therefore entails an 'individual rights-based theory'.[17]

Mill's theory on the essential difference between a state ('mankind collectively') preventing harm to self and preventing harm to others, or public opinion ('mankind individually') discouraging one or the other, which Brock quoted ([8], p. 21) and considered non-utilitarian, follows from an even more pronounced commitment to those same individual rights. That commitment again rests on an elaboration of the very principles introduced in Bentham's theory. His account of individual rights merely adds a further premise and thereby gains an additional conclusion.

Bentham considers individual rights exclusively in terms of everyone pursuing his own private happiness. He appears to believe optimistically that such a libertarian strategy will most effectively promote the greatest happiness of the greatest number. Mill demands not only the freedom of thought, but also of speech, press, and assembly ([15], p. 12), and he does so for two reasons. First, he thinks that happiness for some is found not merely in the private pursuit of pleasure but also on the basis of a generous concern for the condition of society at large. Second, he believes that this concern will have favorable effects in increasing the greatest happiness for the greatest number. The greatest happiness is then no longer merely a sum of all the individuals' happiness to be calculated for each occasion of utilitarian evaluation, but a historical task of human progress requiring occasional reform of encrusted traditions ([16], p. 465). Since such reform can be brought about only by opposition to official *decorum*, established habits and public opinion ([16], pp. 32f, 55f), that same opinion must not be permitted to decide the question as to whether someone is about to harm himself by not pursuing the conventionally accepted happiness or whether he is sacrificing personal pleasures for the sake of humanitarian goals ([15], p. 76). It is this incompetence of the public (in both its forms) which motivates Mill's emphasis on the distinction between harm to self and harm to others and hence his special emphasis on individual rights. It is, in addition and most importantly, the grounding function, which that individual rights-based theory serves for Mill's particular utilitarian criticism directed against public opinion restricting the ways in which non-conformist individuals follow their humanitarian inclinations, that exposes the utilitarian relevance of his harm-to-self versus harm-to-others distinction.

As a result, it appears that the two elements Brock found excluded from Bentham's theory are quite compatible with any particular utilitarian endeavor. The utilitarian nature of Szasz's theory can thus no longer be denied on the basis of it containing these two elements.

(b) The preceding argument has suggested that (1) Bentham and Mill are linked by one and the same utilitarian endeavor (only set in different historical contexts) and (2) Mill's writings present a more thorough understanding of that endeavor. I shall now discuss these suggestions in detail.

(1) Brock's opposition to the first thesis is understandable if we consider that Mill criticizes Bentham's theory and that both thinkers have rather different things to say about human happiness and the like. Nevertheless, it seems undesirable to separate Mill from a philosophical tradition to which he is historically linked. In addition, the inherent unity of that tradition becomes clear if we consider utilitarianism as essentially not a certain theory about the state, ethics, the law and human happiness, but rather as a particular kind of political criticism based on a particular meta-moral position.[18] This is not to deny, obviously, that such criticism requires for its justification underlying theoretical frameworks, which may but do not have to be different for each situation to which the criticism is addressed.

The kind of criticism typical of the utilitarian endeavor thus understood employs the principle of happiness-maximization (or any equivalent principle underlying the meta-moral position) in an *ad hominem* fashion. It is directed against any more or less institutionalized social power that abuses or is threatening to abuse paternalistic reasoning.[19] Such abuse is supposed to consist in a social practice where power is being exerted unchecked, i.e., where the victims, by being deprived of categories for criticism, are denied the means of defending their liberty. That deprivation is effected by the pretense of a concern for the common good insofar as that good encompasses their own, while in effect the profit goes somewhere else: to particular factions or to the stabilization of a societal *status quo* that has outlived its function.

The respective underlying theoretical framework specifies among other concepts the meaning of 'happiness' (as in 'happiness-maximization'), of 'institutionalized social power' (in view of a moral obligation ascribed to that power, which imples that the utilitarian criticism is understood as an *ad hominem* argument), and 'suppressed' (indicating the special historical setting of that argument). The patent differences in Bentham's and Mill's opinions concern historically determined differences in that theoretical framework underlying their equally utilitarian arguments.

Thus, Bentham insists on the pursuit of *private happiness* (understood in terms of refined epicureanism) ([3], pp. 16ff), because he is arguing against a ruling class that pretends to secure a general good on moral grounds. Mill, on the contrary, insists on the possibility of pursuing happiness by *noble altruism* as well, because the public opinion he is confronting restricts the acceptable 'happiness' to that very private epicureanism Bentham had espoused.

With Bentham the institutionalized social power abusing paternalism is represented by the *few governing aristocrats* suppressing the many subjects. In the case of Mill that power is also represented by *public opinion,* or by the many suppressing the few. Nevertheless, Mill can appeal to the principle of happiness-maximum as well as can Bentham, since (i) the happiness of the many is – for both Bentham and Mill – lessened by any injustice done to the few (noble altruists) and since (ii) Mill allows the happiness of the many (or rather the qualitative improvement of that happiness) to be furthered by those generous few in such a way that the many would fail to promote their own advantage were they to discourage the humanitarian zeal of the few.

(2) The more thorough understanding of the utilitarian undertaking as it is found in Mill is distinguished by a level of reflection that allows him to integrate a proper understanding of the general function of utilitarian arguments into the very theoretical framework underlying his particular one. Hence, Mill's account at the same time meets the condition (defined above)

for adequacy in view of grounding the rules by which to evaluate policies of paternalism. It is this result that links my discussion of the meaning of utilitarianism with the subject of the present volume. This result also links that discussion with my thesis that Szasz's arguments, by being interpreted on the model of Mill's utilitarian reasoning, present indeed a criticism that ought to have been here considered.

Bentham's list of human needs, while it was designed to determine the indubitable minimal requirements to be respected by all public policies, could be profitably mistaken to present a sufficient list and be made to serve new policies for the suppression of social criticism as 'irrational' ([25], p. 29). Thus, Bentham's very libertarian thought gave rise to that same narrowly hedonistic determination of rationality that Mill had diagnosed as the tyranny of public opinion ([15], p. 4, [14], p. 15) and that Szasz discovers at the root of an oppressive society today.[20] What had been intended as a means of political protest degenerated into a means of veiling intolerance behind a seemingly uncontroversial 'naturalist' anthropology.[21]

Contrary to that, Mill's theory provides safeguards against the abuse of paternalism. This is not to say that Mill's account of human happiness (in its relation to his view of the state not only as a guarantor of private happiness, but equally as a social educator of mankind)[22], taken by itself, could not become dogmatized in just as oppressive a manner, as can be found in Inquisition reasoning, or in the situation that confronted Brock. Indeed, any particular utilitarian criticism of a particular existing definition of acceptable happiness carries the burden of proving that outside that definition happiness can still exist. Such criticism thus requires commitment to yet another definition that may eventually lend itself (quasi by definition of 'definition') to excluding still further unusual individuals from the realm of respectability. Yet that very possibility is reflected within Mill's theoretical framework, taken as a whole. It is in response to this danger that Mill specifies a quasi meta-utilitarian intention by endorsing not merely a particular idea of happiness that serves its revolutionary function only for the moment, but also the merely *formal ideal of liberty*.

Mill is aware of the fact that public opinion can never be trusted (or at least not before a final stage of moral education has been reached, and it is clear to him that that stage is yet far away). But he is also aware of the fact that the individual must remain protected from government imposition ([15], p. 112), even where that government has assumed the very educational function Mill devised for it. The meta-utilitarian endorsal of liberty in the formal sense by which – within his theoretical framework – Mill supplements his particular utilitarian criticism, accounts for the fact that the task of happiness-maximization, once understood as a historical task, requires political liberties not just for the reformers of the particular situation at hand, but for critical minds in general. His very theory, though grounding an argument that is directed against prevalent dogmatization of Bentham's critical ideals, allows for another Bentham to recur, once his own (Mill's) more ambitious ideals have assumed an oppressive function of their own.

3. *Utilitarian Resolution of the Inconsistencies Attending Szasz's Theory*

The understanding of utilitarianism outlined in the preceding section grew out of an analysis of certain *ad hominem* arguments (against self-interested abuse of paternalistic reasoning) in which utilitarian criticism was employed. It is the very kind of argument analyzed there, which can be found exemplified by the reasoning reconstructed for Szasz's antipaternalistic intentions in the first part of this paper. Whereas Brock had reached his negative conclusions concerning present mental health policies on the basis of the epistemological difficulties of singling out the cases to which his criteria apply, the essentially political nature of Szasz's concern could be accounted for within its utilitarian interpretation.

On the basis of the reflective level of critical utilitarianism as presented in Mill's theory, it now becomes possible to resolve the inconsistencies that have been found to hamper Szasz's position. As a point of departure to that resolution, a likely objection to the very possiblity of interpreting Szasz in Mill's terms may be considered. Earlier it was shown that neither the moral desert theory of legal punishment nor the emphasis on individual rights as conceived in Mill's arguments could repudiate the structural similarity of Mill's to Bentham's utilitarian criticism. Nevertheless, it might still be thought that the kinds of individual rights and moral deserts admitted by Mill are incommensurable with those kinds figuring in Szasz's account. (1) Mill's emphasis on individual rights has been illustrated by his distinction between harm to self and harm to others. That distinction was drawn in view of altruistic noble souls who refuse to act in a narrowly self-interested way. Yet, Szasz's emphasis on individual rights pertains not to great reformers but to simple fools. (2) Moreover, Mill's desert theory of legal punishment served the purpose of establishing a unity between utilitarian considerations and the concerns of justice, whereas for Szasz that same theory served the purpose (among others) of establishing a strange unity of utilitarian consideration and the scandals of injustice.

Against these objections, it can be shown that Szasz is indeed one of those "future Mills" whom Mill had envisaged when devising his meta-utilitarian argument. The following parallels make that clear:

(i) Just as Mill – if he wanted to oppose the paternalistic interference against which his criticism was directed – had to make sure (within his theory) that altruism would be counted among the acceptable sources of happiness and not among the motives of harming oneself; so Szasz – in view of the same critical purpose – has to make sure that deviance be counted among the acknowledged (if unwise) means of pursuing happiness and not among the conditions leading to harming oneself (*simpliciter*).

(ii) Just as Mill was motivated in this by his goal to have great men be given free range for fulfilling their humanitarian task and thus contribute to the growth of general happiness, so Szasz is motivated (among other reasons) by his goal to have behavioral deviants be given free range for fulfilling their humanitarian function of exposing social misery in such a way as to motivate Mill's great men for their humanitarian tasks.

(iii) Just as Mill took pains to define altruism as belonging to the kind of happiness Bentham had prescribed in order that his argument (for tolerating the pursuit of altruistic designs) acquire *ad hominem* force with the Benthamites, so Szasz takes pains to decipher deviance as an indirect way of pursuing happiness under the conditions of social victimization in view of stimulating the humanitarian projects prescribed by Mill, in order that his argument (for granting liberty to the insane) acquire *ad hominem* force with the Millians.

(iv) Just as Mill's emphasis on the fact that societal injustice is engendered by encrusted habits and requires occasional reform derived from his observation that Bentham's sum of the many egoistic pursuits cannot be relied upon to secure the greatest happiness for all, so Szasz's emphasis on the fact that societal injustice is engendered by the apologetic use of the medical model and requires occasional exposure of scandal may derive from his observation that Mill's liberties granted to potential reformers cannot by themselves be relied upon to effect the needed reforms. Hence, it is Mill's view of maximal human happiness as a historical task that renders intelligible the purgative utilitarian function of scandal (3rd difficulty)[23] and thus explains the difference between Mill's concern for both (tolerating) great men and (accounting) for justice versus Szasz's concern for both (liberating) fools and (exposing) injustice.

The nature of that injustice is clarified by Szasz's double account of deviance (1st difficulty), which in turn becomes intelligible when seen in relation to Mill's view of the state as a moral educator of mankind. According to that view the state is not only obliged to provide the boundary conditions for the private pursuit of happiness, but also entrusted with actively improving the quality of that happiness. Szasz's double account of deviance mirrors the failing of the state in both respects: People become behaviorally deviant because they are victims of social injustice (failure of allowing for each to pursue his happiness), and the medical model, designed to hide this failure, assumes the further 'educational' function of demoralizing[24] those who will take unfair advantage of the social loophole offered them (failure of improving the subjects). While Mill is opposed to social and political suppression because it hinders the free development of moral impulse (and thus hinders the maximization of happiness), Szasz opposes such suppression because it furthers the free development of immoral impulse (and thus presents a similar obstacle). Finally, the nature of that additional moral function of the state, which explains Szasz's double account of deviance as serving a double political criticism, is clarified by Szasz's double account of personhood (2nd difficulty), which in turn becomes intelligible when seen in terms of another set of parallels to Mill.

I shall first consider that double account of personhood and only later draw the conclusions regarding the state. The parallels are:

(v) Just as Mill has a meta-utilitarian argument demanding that liberties should always be granted in a purely formal sense, and that no criteria defining acceptable happiness are ultimately valid; so Szasz (while considering yet another dimension of the happiness-criteria for defining rationality and personhood)[25] has a meta-utilitarian argument demanding that personhood should always be granted in a purely *a priori* sense, and that no criteria defining acceptable personhood are ultimately valid.

(vi) And just as Mill derived his meta-level utilitarian argument from a reflective understanding of his particular utilitarian criticism ("happiness is not only found in egoistic pursuits, but also in altruism") so Szasz derives his meta-level utilitarian argument from a corresponding understanding of his corresponding criticism ("rationality is not only found in using means to ends consciously and directly, but also unconsciously and indirectly").

Taken in this sense, the fact that Szasz has a double account of personhood mirrors the doubleness of employing a utilitarian argument (used for criticism of the immediate situation at hand) and supplementing it by a meta-utilitarian argument (indicating a direction for future such criticism).

The specific kind of his double account becomes intelligible if we consider the specific way in which Mill's theory also accounts for his respective doubleness of utilitarian and meta-utilitarian arguments. The maximization of happiness can only be pursued in the process of gradually improving the quality of that happiness. This task cannot be completed at once, but requires historical progress, or a series of ever new particular social institutions ([14], p. 16; [16], p. 258f) developing ever new suppressive mechanisms, but driven toward perfection by ever new 'Mills' devising particular criticisms according to Mill's meta-utilitarian maxim. As a result, the state in both its functions is seen by that theory in a twofold manner: In its ideal form as envisaged by the meta-utilitarian argument and in its historical manifestations as addressed by the utilitarian criticisms. That criticism can claim *ad hominem* force only insofar as the particular state criticized can be assumed to see itself as called towards its own ideal form.

On the basis of these considerations it now becomes possible to use Szasz's double account of personhood for clarifying that theoretical view of the state, in particular, of its moral function and in view of the twofold manner just indicated. For Mill, the state is a moral educator insofar as it endeavors to shape its subjects; for Szasz that shaping will always have to take place according to a definite notion of personhood, with the educating state at the same time understanding that its own perfection depends on those subjects of its education ever re-shaping that very notion which it then in turn applies. Its 'education' will therefore consist not in a simple conditioning, but – on the basis of ever new historical understandings of 'personhood' – in an appeal to what is universally 'personable'. Thus to return to Szasz, the state must be argued to have the moral obligation of bringing about the personhood of its subjects by treating them as persons precisely insofar as that state is at the same time obliged to conceive those subjects as constituting what personhood means in the first place. The relation thus encompassing that doubleness of causing and of accepting a pre-existent personhood, which presented Szasz's second difficulty, can be conceived to make sense as a relation of *acknowledgement,* implicitly contained in Mill's understanding of the state.

With this resolution of the inconsistencies attending Szasz's position in view of an understanding of Mill's reflective account, the utilitarian reconstruction of Szasz's reasoning has come to an end.

It has by now become obvious that the effort to provide such a reconstruction has entailed both that violence be done to Szasz's writings (by leaving out most of what he said and by changing the relevance of much that remains) and that the meaning of 'utilitarian' was stretched beyond, and at the same time restricted within, the range of its usual signification. Hence the defense of my thesis (that Szasz's theory must be so reconstructed) can appear plausible under one or the other of two conditions: Either that reconstruction has so drawn his theory into a more favorable light that one is willing to tolerate the rather odd interpretation of 'utilitarian' for the sake of its underlying that exposition; or that understanding of 'utilitarianism' has seemed so suggestive for a general interpretation of Mill or Bentham that one is willing to tolerate the new respectability of Szasz's rather odd theory, because it follows from such an understanding.

In any event, the direction for future discussion of mental health and its public policy implications that was to follow from my interpretation of Szasz's reasoning can now be indicated: An adequate treatment of the problems involved in the issue requires that one does not restrict one's attention to the particular political, social and medical frameworks presently given and thus to the task of justifying associated policies or of criticizing only special aspects of their application. Rather, the justifiability of those very frameworks should be scrutinized in the light of their implications for behavioral deviants. The significance of Szasz's criticism lies in the fact that he has applied the utilitarian principle of happiness-maximization (in Mill's reflective version as implying the respect for everyone's liberty) to a group of human beings who were previously excluded from such considerations. He has placed the 'mentally ill' in a class with the socially disadvantaged. If we take policy problems to concern how under given conditions one should reach desired goals, then Szasz's alerting us to unexamined assumptions implicit both in our assessment of those conditions and in our view of the results has opened the way for much more radical questioning. He has raised what in this volume merely appears to be a problem of policy to the level of a political problem.

In addition, much of the psychosomatic medicine of recent years has suggested that many so-called physical diseases have an intimate connection with those very 'problems in living' Szasz diagnoses as underlying psychiatric 'illness'. Considering all the intermediate arguments of Szasz, one would reach the conclusion that his criticism of institutionalized psychiatry as an occasion for abuse of paternalism would commit him to a corresponding criticism not only of the present welfare system but also of institutionalized medicine in general.

As a consequence, Szasz would have to entertain a rather encompassing notion of social victimization, since there would be few individuals not victimized by society. How could one weigh the decrease of suffering that could reasonably be expected against the loss of personal liberty solutions might bring about? However, Szasz also insists upon individual, not only social, efforts to alleviate the sufferings of others. He reopens the 'classic' utilitarian question: At what point does a course of action taken by the government for the sake of the common good become counterproductive?

It is in view of this network of questions that positive or constitutive theories such as the ones by Rawls and Nozick are needed in order to establish a rationale for selected ways of reasoning in these regards. It is in the context of such encompassing theories that the issue of mental health should be discussed again.

Department of Philosophy
Pennsylvania State University

NOTES

* I wish to thank Michael Hayes for his painstaking efforts in reducing Germanicisms that constitutionally infect my vocabulary, grammar, style, and general frame of mind.

1 M. S. Moore has published a very thoughtful critique [18] of the allegedly philosophical presuppositions on which Szasz sometimes claims to rest his arguments against 'mental illness'. Szasz's polemical preoccupation unfortunately causes him to support his humanitarian goals by all too heterogeneous sets of *argumenta ad hominem*. In order to extract assent from his American audience, he appeals to conceptual commonplaces derived from prevalent college teaching ideology supposedly imbibed by every undergraduate. His arguments, designed for the unthinking, lend themselves to easy refutation by Moore.

Yet it seems to me that this is not the most profitable manner of responding to the challenge presented in Szasz's work. Regardless of all their philosophical flaws, his reasonings have political implications that seem to present his most valuable and at the same time most tenable contributions. For the sake of attending to these contributions, my presentation of Szasz's position will simply disregard those arguments that are obviously unwise.

2 By suggesting that a utilitarian argument is generally assumed to be a 'respectable argument', I do not wish to imply that it is generally considered valid nor even that I myself consider it valid. I am using the utilitarian interpretation for exactly three reasons: (1) in order to give some structure to Szasz's reasoning that renders it philosophically more intelligible by fitting it into available categories; (2) because such an interpretation is particularly apt for explicating what is relevant in Szasz's theory for political thought; and (3) because the specifically critical meaning that I shall assign to the term 'utilitarian' may even contribute to more philosophers finding utilitarian arguments valid or at least finding less cause for objection with them.

3 In view of this argument it seems to be quite superfluous that Szasz takes so much pain in proving mental illness not to be a disease, and in committing himself to a questionable understanding of somatic medicine as merely dealing with flaws in the body-machine. It would be quite sufficient for him to argue that problems in living, while they *respond* to drugs (and while the prescription of drugs may in some cases be

even necessary), are not adequately accounted for in view of such a treatment alone.

As a consequence, Moore's responding to Szasz's unfortunate imputation of a category mistake to the common understanding of 'mental illness', by insisting on the possibility to correlate physical and mental events [18], is misleading, since it hides the more important question: How much is gained by establishing such a correlation?

4 In a public health care system, such care will be entrusted to government agencies, educational and military authorities, and to particular employers who wish to contribute their share of social services. It is these institutions who employ, pay, and empower psychiatrists to treat their patients, at times even without or contrary to their consent. Social institutions must be supposed to serve the interest of society – nor is any contract established between patient and physician by which to balance that bias. Hence, no sufficient safeguards are provided to ensure confidentiality and to protect the interests of the patient, both of which are indispensable for that atmosphere of confidence without which no therapy can be successful.

Of course, the same holds for institutional physical health care. But since the values underlying the categories of physical illness are more basic 'prudential' values, they are also more stable and less vulnerable to the particular ideological leanings of a given society. The values underlying behavioral norms, on the other hand, are vastly more flexible and lend themselves more easily to political abuse – a distinction Margolis fails to take seriously enough for the issue in question.

5 A further argument by Margolis against Szasz's thesis of a difference in kind between physical and mental illness focuses on various species of mental illness, and various degrees of similarity between their symptoms and those of physical illness. Margolis concludes that the issue of classification must be decided not in a rigorous but in a flexible manner and in view of the contingent phenomena in each case. Considering Szasz's insistence on the communicative function of mental illness, this argument by Margolis must appear as confusing, as would be the analogous claim that some species of signs (for instance: hieroglyphs) should be classified as objects, because they are rather more similar to them than to other species of signs.

6 In quite a similar vein John Stuart Mill, when considering the phenomena connected with 'folly', "lowness or depravation of taste" ([15], p. 77), is confronted with two possible interpretations. If one considers them merely as breaches of duties to self, they remain a private affair. If one considers the dispositions underlying that conduct and the far-reaching societal consequences of the existence of such dispositions, however, then the matter becomes 'political'. Mill's endorsement of the first interpretation is motivated by the danger of abuse inherent in the legal measures, which danger would necessarily ensue from viewing such conduct as 'political'. The cost of potential abuse would far outweigh the benefit expected from an enforcement of self-culture. Hence, for Mill just as for Szasz, classificatory decisions are undertaken in view of the social consequences and thus become subject to the utilitarian principle.

7 The only one who does without criteria is Sartorius [23], but it is not clear whether this is for him a matter of principle for which he would be prepared to argue, or if the scope of his paper simply did not permit him to argue adequately for those criteria he would endorse, so that he preferred not to mention them at all.

8 Even where legal definitions focus on 'responsibility', the latter term is understood as 'rationality of conduct' [19] or as 'freedom', which in turn is a capacity to be motivated by 'good reasons'. Coleman [9], in his discussion of liability in tort law, tries

to justify morally why mental illness precludes the imposition of recompense only in those cases where the minimal conditions for agency are not satisfied. Yet he is not quite clear about the consequences of such a case. Sometimes he says that the lack of agency precludes *fault*. But as fault was defined in view of merely a causal relation between certain kinds of events, it is quite incomprehensible why, in the second part of the argument, the additional qualification of having to be an *action* is added to the antecedent-event. Instead of arguing, as Coleman does, that victims of floods, etc. have no right to recompense because no *agency* was involved, one might equally argue that in such cases no person is at fault, i.e., that already the weaker condition is not satisfied. Similarly, if a person was forced by someone else to harm a third, fault can be excluded from the person immediately causing the harm not only because he did not *act* (as Coleman argues), but can be excluded from that person because it can be attributed to the one who forced him, and who thus is responsible for the remote but more 'momentous' cause of the harm. Sometimes, on the one hand, Coleman says that the lack of agency precludes *liability*. But as the imputation of liability proceeds independently of subjective factors characterizing the situation of the one causing harm, and as 'action' is defined by Coleman in terms of such very subjective factors (antecedent intentions, and the capacity of translating them into bodily movements) as have been excluded from consideration in judgments on liability, it is hard to understand why nonagency should be relevant to liability-issues. In other words, Coleman has justified the existing tort law by a regard to the victim; it is incomprehensible, therefore, on the ground of his own categories, why this moral basis for the law should suddenly be irrelevant in cases of harm due to non-action, where the victim suffers just as much. Or the necessity of distinguishing between agents and non-agents is not itself morally justified.

9 Even Margolis's 'prudential values' [13] are relativized by so many qualifications that they do not repudiate this fact. Nor is it relevant to the issue of rationality as raised by Szasz that we, in fact, all agree on some basic intuitions concerning rationality (or agency, for that matter), and that this agreement informs much of our practical decisions in everyday life and intellectual transactions [15]. What constitutes sufficient probability concerning rationality may yet be insufficient to safeguard against undesirable political consequences of such intuitions. Szasz therefore enjoins us not to *compare* our intuitions with his views, but to *change* the former in accordance with the latter.

10 Szasz claims that "moral . . . and . . . legal sanctions . . . are more dignified and less injurious to the human spirit than the quasi-medical psychiatric sanction of involuntary mental hospitalization" ([25], p. 119).

11 One might feel tempted to evade that difficulty by counting his arguments on the desirability of scandal with all those other Szaszian utterances that have to be disregarded if his theory is to make sense. This would leave us the alternative of placing the thrust of Szasz's utilitarian argument against the insanity defense on the desirability of 'therapeutic punishment' ([24], p. 213ff) instead. As a consequence, even if the alternative between excusing-and-committing and punishing would not exhaust all options, there would be an independent justification for the latter policy on the basis of its therapeutic effect on the deviant delinquent.

Yet it quickly appears that nothing is won by this strategy, since any account of that therapeutic punishment would mean reintroducing into the argument the scandal we just tried to avoid. Concerning the manner in which such a policy would work, the alternatives open to our understanding the property of personhood (issue two) have to be considered. Either that policy would work like an act performed by society on the delinquent, that is, in the sense of a therapy mimicing punishment, or it would work like an appeal to conscience and a sense of responsibility that exists in the delinquent even though he had weighed values in an unwise manner ([24], p. 115), that is, in the sense of punishment that is hoped to effect a metaphorical 'therapy'. The first alternative corresponds to personhood as a property that must be ascribed necessarily to people for moral reasons, the second to personhood as a property that must be ascribed only where the facts justify it. Concerning the manner in which that alternative will re-introduce the desirability of scandal into the utilitarian argument of the third issue, we have to consider as well that the utilitarian argument in the first issue rested on the victim-of-society account for behavioral deviance. (The conflicting moral-deficiency account will therefore have to be reconciled with that first account, if it is to make utilitarian sense.) As a result, if therapeutic punishment is essentially punishment, then that policy would imply that some one is punished by society for having become the victim of society. If, on the other hand, therapeutic punishment is essentially a therapy (as an act by which to transform the delinquent deviant into a person), then that policy would imply that the state can avoid doing that delinquent the injustice of not treating him as though he were a person (and thereby of taking away his chance to become a person) only by doing him the injustice of punishing him unjustly.

12 According to Inquisition reasoning, the witch's capacity to 'form purposes' (e.g. to avoid being burnt alive), to 'weigh alternatives' (e.g. to confess her crime or to remain silent), and to 'act on the result of that deliberation' ([8], p. 26) is 'defective'. Either she labors under a 'defect of reason' (e.g., the devil deludes her with the false promise that he will save her from the scaffold), and it is impossible to reduce her 'decision' to 'differences in values' ([8], p. 29) (since 'going to hell' is the negation of all values). Or she labors under a 'weakness of will' ([8], p. 26) (e.g., the devil has induced taciturnity) and is thus unable to act as she believes she ought to. The Inquisitors believe that she will be better off if treated paternalistically (e.g., is tortured to the degree that the physical pain will overcome the devil's machinations). They have "reasons to believe that the person is not able to act as (s)he believes (s)he ought . . . to act" ([8], p. 27), because they have studied demonology. They know that her "long term desires are in conformance" with that action though the present ones are not, since any rational being (devils excepted) would prefer to be tortured a little if he may thus escape the more serious harm of eternal hellfire.

(Samples of such reasonings can be found in ([6], Book IV, Chapter II and III; "De probationibus ad evincendum magiae crimen necessariis, De voluntaria et coacta magorum confessione", and ([10], Book V, Section IX; "De quaestione, seu tortura".)

13 Out of the very complex pattern of Bentham's motivations I am isolating here that train of thought relevant to the issue under discussion: the relationship between utility and morality. Yet this aspect of Bentham's theory is also crucial for understanding why Bentham would derive his 'principle of utility' from the thought of David Hume ([1], p. 242). Apart from other reasons motivating his attempt to reform the law, such as dissatisfaction with a confused administration, with an unintelligible legal theory and with the fiction of a "social contract", as with an unsatisfactory way of backing the defense of liberal ideas, Bentham shares with Hume a general interest in defining which kinds of reasoning can legitimately be employed in political arguments, and what are the undeniable principles of human nature on which these reasonings ought to be based.

It is on these grounds that Bentham sympathizes with, for example, Hume's indignation at the public enforcement of religious and moral principles (see in particular [4]) and on which Bentham thus undertakes to determine the limits within which a government, in following its obligation to promote the common good, is entitled to interfere legally with the liberty of the citizens.

14 Bentham himself states that utility (justice) and private morality have identical goals ([3], p. 142ff). But this should not be taken to mean that moral and utilitarian prescriptions cover the same ranges of decisions. Rather, the principle of utility provides a wider context in which moral concerns (in the narrow or private sense of the word) have their place. Moral norms regulate individual conduct. But the question concerning which aspects of that conduct should also be regulated by the law cannot be answered on the grounds of moral norms but only in view of the utility such regulation would have for that common goal. Thus cases principally 'unmeet for punishment', because unmeet for institutional action, are those where a legal enforcement of the goal of happiness for everyone would, in effect, create greater evils (calculating the pains of deterrence and expectations as experienced by every individual) than the toleraton of certain actions not conducive to that goal. They are also cases where the same effect of negatively enforcing certain actions could be realized 'more cheaply', i.e., by non-institutional sanctions of public opinion. As a consequence, Bentham's exemption of particular cases from punishment (for example, where the victim consented, or in cases of *ex post facto* laws), rather than violating the utilitarian principle, as Brock believes ([7], p. 253f) agrees with it.

15 Brock's tenet that "values do no real work in utilitarianism" ([8], p. 20) because they become derivative from a regard for consequences superimposed on those values is therefore misleading. On the contrary, the evaluation of such consequences depends precisely on the values realized in them. If values are goods, goods are desired, and the attainment of a desire constitutes happiness ([16], p. 237), then the realization of values is defined as human happiness and *vice versa*. Alleged moral values attending a course of action not 'cashed in' in terms of happiness are to be rejected as unreal. The utilitarian principle establishes a moral meta-language permitting scrutiny both of the use of moral terms in view of the values appealed to and of the significance of those values in view of that happiness they ought to promote.

16 Mill misunderstands this critical intention of Bentham's theory or else he addresses only the subsequent dogmatization of those critical principles when he reproaches Bentham for doing injustice to the more generous minds and for debasing the human race ([14], p. 12).

17 No reward for public utility could justify the sacrifice of individual rights. Any advantage incurred in such a policy would be outweighed by the pains of benevolence (pity), of imagination, and of expectation of similar dangers to one's own rights, all of which would be experienced by those other citizens who allegedly profit from it ([2], p. 18). It is characteristic of the utilitarian optimism that conflicts between the egoistic and the benevolent feelings of individuals are always presumed to guarantee that the advantage of a majority (in terms of narrow expediency) cannot outweigh the 'moral' costs of hurting a minority.

18 By suggesting this understanding of utilitarianism that, within the scope of the present essay, must remain all too sketchy and should be considered a thought-experiment rather than a ready theory, I am not only trying to render intelligible the

common ground on which both Bentham's 'individualist' and Mill's 'social' utilitarianism ([11], Vol. 3, p. 374ff) stand, and to account for the fact that utilitarian reasoning has been found compatible with ethical positions as diverse as are for example naturalism and non-naturalism ([17], p. 444ff). Rather, I am also trying to indicate a way in which the utilitarian undertaking in the narrow sense of a particular kind of political criticism can be defended against some of the criticisms analyzed by Brock [7]. In particular, I want to suggest that utilitarian reasoning in the sense outlined here can be profitably used for criticism within particular political situations, even though the underlying conceptual framework does not meet the requirement of providing a theory of the state that determines in general the rules of justice that should guide its actions. It is, after all, this failure of utilitarian theories (as they are commonly understood) to provide a positive, or constitutive theory of justice that underlies Nozick's criticism of utilitarianism. Yet, criticisms of a particular unjustified paternalism can be derived already from premises that would not suffice for criticism of all possible such paternalisms, let alone of all possible injustice. As will be seen later, Mill's utilitarianism can be interpreted in such a way that it indeed provides for all possible future criticism (at least of unjustified paternalism). But it will also be seen that his theory does so not 'materially' on the basis of categories laid down once and forever, but 'formally', by furnishing an ideal to which ever new criticism can refer in terms of ever new historical understandings of the significance of that ideal. While Nozick wishes to present one conceptual network holding all possible solutions, Mill offers a rule by which to generate ever new solutions.

19 Mill's criticism is directed both against unjust interference from the side of the state (arising from a failure to abide by the utilitarian moral commitment for a general good, and hence arising by definition from some particular self-interest) and unjust interference from the side of public opinion (arising either from general indolence, or from the intolerance of custom); compare for example ([12], p. 117ff). Yet, I would like to argue that his specifically utilitarian criticism addresses only the state, or, more specifically, the view that those representing the state, or those in power of government, should have concerning the state as well as concerning the specific function of public opinion. The kind of criticism by which he addresses public opinion directly should not be called utilitarian since it cannot function in an *ad hominem* way. That is, it cannot appeal to any self-understanding in the first place. This kind of criticism should be called pedagogic rather than utilitarian.

20 It is perhaps not accidental that Bentham, even though he is quite aware of the arbitrariness of any criterion by which one could distinguish those in need of a ward ([3], sub. XLIV, p. 125), has no misgivings in defining the duty of such a ward as the government of some individual "precisely in the manner in which this latter" in point of prudence "ought to govern himself" (*ibid.*, sub XLVI). Bentham was not so much concerned with scrutinizing the grounds on which someone was declared insane, as with devising ingenious techniques for their effective supervision; see [2].

21 "(1) laws and social arrangements should place the happiness of every individual as nearly as possible in harmony with the interest of the whole . . . (2) education and opinion . . . should so use that power to establish in the mind of every individual an indissoluble association between his own happiness and the good of the whole" ([15], p. 418).

22 Such a consequence would not arise, as Brock would have it, on the basis of

preferring public expediency over individual rights, but because, again, an originally critical account of human happiness has become dogmatized.

[23] Of course, an act of demoralizing someone is not at all like an act of, for example, dishonoring someone, nor even like depersonalizing. It is not an act performed on someone else simply but rather it consists in providing an opportunity that makes it all too easy for the other to demoralize himself. If is only because of this peculiar way in which a state can be at fault for demoralizing its subjects who are willing to let themselves be demoralized that the utilitarian justification of the punishment of the insane in view of the scandal thereby created becomes plausible: even disregarding the utility of scandal that punishment is just with regard to the delinquent himself.

[24] Thus Szasz blames institutionalized psychiatry for creating legal loopholes for those who would otherwise suffer from unjust laws. By such cooperation, he argues, psychiatrists contribute to the perpetration of disastrous policies, instead of allowing their injustice to be brought into the open ([25], p. 90, 96f).

[25] Sartorius [23] seems to suggest that Mill himself would have consented to the inclusion of the mentally ill in the class of those individuals whose greatest happiness implies their liberty. Therefore, he quotes Mill as excepting only children and primitives from having a just claim to individual freedom. He omits Mill's additional qualification according to which his argument applies to only those of 'mature faculties'. Whether for Mill this would have excluded 'ordinary' mental patients (as distinguished from the retarded or mentally deficient) is hard to determine. On the one hand, Mill is remarkably more willing to consider 'eccentrics' as persons in the full sense of the term than his contemporaries. On the other hand, he holds that no one should be free to be not free ([15], p. 104), and in his critique of Bentham [14] there is no indication of dissatisfaction with the latter's somewhat paternalistic attitude towards the mentally ill.

BIBLIOGRAPHY

1. Bentham, J.: 1962, 'A Fragment on Government,' in *The Works of Jeremy Bentham,* Vol. I, ed. by J. Bowring, Russel and Russel, New York, pp. 221–297.
2. Bentham, J.: 1962, 'Panopticum, or the Inspection House,' *op. cit.,* Vol. IV, pp. 37–173.
3. Bentham, J.: 1962, 'An Introduction to the Principles of Morals and Legislation,' *op. cit.,* Vol I, pp. 1–155.
4. Bentham, J.: 1962, 'Swear not at all,' *op. cit.,* Vol. V, pp. 187–231.
5. Bentham, J.: 1962, 'The Constitutional Code,' *op. cit.,* Vol. IX.
6. Bodin, J.: 1603, *De Magorum Daemonomania, Seu Detestando Lamiarum Ac Magorum cum Satanae Commercio Libri,* Wolffgang Richter, Frankfurt, Germany.
7. Brock, D.: 1973, 'Recent Work in Utilitarianism,' *American Philosophical Quarterly* 10, 241–276.
8. Brock, D.: 1980, 'Involuntary Civil Commitment: The Moral Issues,' this volume, pp. 147–173.
9. Coleman, J.: 1980, 'Mental Abnormality, Personal Responsibility and Tort Liability,' this volume, pp. 107–133.
10. Delrio, M.: 1603, *Disquisitionum Magicarum Libri Sex,* Joannes Albinus, Mainz, Germany.

11. Eisler, F.: 1924, *Woerterbuch der philosophischen Begriffe*, Berlin.
12. Halliday, R.J.: 1976, *John Stuart Mill*, George Allen and Unwin Ltd., London.
13. Margolis, J.: 1980, 'The Concept of Mental Illness: A Philosophical Examination', this volume, pp. 3–23.
14. Mill, J.S.: 1969, 'Remarks on Bentham's Philosophy', *Collected Works, John Stuart Mill*, Vol. X, Toronto University Press, Toronto, pp. 3–19.
15. Mill, J.S.: 1947, *On Liberty*, Appleton Century Crafts, New York.
16. Mill, J.S.: 1969, 'Utilitarianism', *Collected Works, ed. cit.*, pp. 203–261.
17. Monro, D.H.: 1973, 'Utilitarianism', in P. P. Weiner (ed.), *Dictionary of the History of Ideas*, Vol. V, Scribner, New York, pp. 444–449.
18. Moore, M.S.: 1975, 'Some Myths about "Mental Illness"', *Archives of General Psychiatry* **32**, 148–97.
19. Moore, M.S.: 1980, 'Legal Conceptions of Mental Illness', this volume, pp. 25–69.
20. Neu, J.: 1980, 'Minds on Trial', this volume, pp. 73–105.
21. Nozick, R.: 1974, *Anarchy, State and Utopia*, Basil Blackwell, Oxford.
22. Roberts, D.: 1974, 'Jeremy Bentham and the Victorian Administration', in *Jeremy Bentham, Ten Critical Essays*, ed. by B. Parekh, Frank Cass, London, pp. 187–204.
23. Sartorius, R.: 1980, 'Paternalistic Grounds for Involuntary Civil Commitment: A Utilitarian Perspective', this volume, pp. 137–145.
24. Szasz, T. S.: 1961, *The Myth of Mental Illness*, Hoeber-Harper, New York.
25. Szasz, T. S.: 1970, *Ideology and Insanity*, Doubleday and Co., New York.

SECTION V

CRITICAL COMMENTARIES

D. L. CRESON

FUNCTION OF MENTAL HEALTH CODES IN RELATION TO THE CRIMINAL JUSTICE SYSTEM

Mental health codes and criminal justice systems are formalized conventions in this society, each providing a means for removing individuals from active participation in the social life of the community. Much has been made of apparent ethical and practical inconsistencies in the two conventions. The procedures prescribed in the two conventions differ in significant ways while the ends achieved in both are, in some ways, similar. It is unlikely that two such well-established conventions with functional similarities do not have a functional relationship in the social structure of which they are parts. If such a relationship exists, then attempts at reform in either convention may have unforseen implications in the functioning of the other convention. It seems probable that an understanding of an existing relationship between mental health codes and the criminal justice system is best approached through a comparison of similarities and differences in relation to social-cultural values that provide the fundamental assumptions on which the social structure is based. The apparent redundancy in the two conventions for extruding deviant individuals from active participation in the social interactions of the community can be explained as necessary, if egalitarian values are to be maintained as key social assumptions. To explore such a conclusion it is important first to compare the differences and similarities in the two conventions.

The two conventions are similar in that both are formal and codified as part of the law. Both are mediated through the courts where a decision is reached as to whether a given individual is to be removed from the community against his or her will. The decision in both cases is made by a judge or magistrate and in some states a jury may play a part in a decision under either convention. The basis for a decision to remove an individual from the social life of a community in either convention is, at least in part, justified in terms of danger, physical or material, to the orderly functions of community life. Such a basis for removing a mentally ill individual is made overt in the ubiquitous phrase in state mental health codes that prescribe involuntary commitment because an individual poses a threat to himself or others. The shared legal and judiciary mechanisms of the two conventions, as well as a similar rationale for removing a troublesome individual – a danger in

B. A. Brody and H. Tristram Engelhardt, Jr. (eds.), Mental Illness: Law and Public Policy. 211–219.

the ordinary social context of daily life – would seem from a functional point of view redundant and inefficient in a highly complex social system, unless the differences in the two conventions provide a rationale for the apparent redundancy.

There are two significant ways in which the conventions differ. First, the various mental health codes purport to deal with a very different population than the criminal justice system – and for different reasons. Second, the two conventions are significantly different in the judicial rules that must be followed in reaching a decision to remove an individual from his or her community. In considering the former, the apparent differences in populations and rationales can be dealt with at the same time in that they are not easily separated.

The various mental health codes are ostensibly directed at a population of mentally ill individuals, while the criminal justice system is directed at a population consisting of individuals who have committed criminal acts. The first promises treatment while the latter prescribes punishment and, only incidentally and occasionally, rehabilitation. There is another difference in the rationales underlying the two conventions. Mental health codes, unlike the criminal justice system, are mediated on the basis of an assumption, both judicial and popular, that removal of a mentally ill individual is for his or her own good.

Differences in populations and rationales might be sufficient to explain the apparent redundancy of the two conventions in an egalitarian society, were it not for discrepancies between the stated goals of mental health codes and the actual effect of involuntary commitment. Mental health codes prescribe a loss of individual freedom in order to provide treatment for a mental affliction. This loss of freedom may occur whether or not a criminal act has occurred. Treatment, however, is not an inevitable consequence of the loss of freedom despite the social rationalization for institutionalization. Goffman ([5], p. 352) among many others has noted the importance of the custodial role of state mental hospitals where most involuntarily committed patients are incarcerated. He writes:

> If all the mental hospitals in a given region were emptied and closed down today, tomorrow relatives, police and judges would raise a clamor for new ones; and these true clients of the mental hospital would demand an institution to satisfy their need ([5], p. 384).

There is increasing evidence that institutionalization may in fact impede efforts to treat the mentally ill ([12], pp. 219–238). The community mental health movement is predicated on just such an assumption, and recent class action suits brought against state mental hospitals have marshalled impressive psychiatric and psychological support for such a position. The efficacy of mental hospitals is brought into further question when it is realized that normalcy may be difficult to recognize in hospital patient populations ([10], pp. 250–258).

If, then, a custodial function is important in state hospitals and treatment is not facilitated, but rather impeded by involuntary commitment to such an institution, a functional explanation for the apparent redundancy in the two conventions must be sought elsewhere. Before considering the second difference between mental health codes and the criminal justice system, it will be helpful to explore briefly something of what is known about deviance and the means of dealing with it in various social-cultural systems.

Deviance is a term used in the behavioral sciences to describe human behavior that deviates from the expected and desired within structured social transactions. Social theorists such as Durkheim, Parsons, and Merton conceptualize deviance in structural terms ([7], p. 4). Merton ([8], p. 133) explains deviance on the basis of a discontinuity in an individual between values of his social cosmology and the day-to-day relationships he must maintain while operating within a social context. Whatever the cause of deviance, it is present in all social-cultural systems and inevitably sets into motion modes of social control that at least theoretically act to reduce or eliminate the deviant behavior ([2], p. 48). Most of these modes are informal and frequently covert, for example, signs of disapproval, passive resistance, and failure to respond.

The rules that govern social interactions in a given social system define what is deviant and what is not. These rules, taken together, have been referred to as an 'ethical code'. "The effect of ethical codes", according to Leslie White, "is to produce unity and harmony of human conduct in the interest of the system as a whole" ([11], p. 39). Social mechanisms for enforcing rules of etiquette act in the service of efficiency and harmony in the social system in which they occur.

All social-cultural systems rely heavily on informal sanctions, but most also provide formal, even ritualistic mechanisms for dealing with deviant behavior. These formal mechanisms can be conceptualized as 'last resorts' in continuing efforts to insure the efficiency and harmony of the system. Such formal mechanisms may be divided into two broad categories, those that provide a means within the ongoing social interaction of the community for dealing with deviance and those that either temporarily or permanently extrude deviant individuals from the community.

There are many mechanisms that have been developed in diverse social systems for dealing, in a formal manner, with deviance within existing social interaction. Payments in cash or goods for transgressions ([4], p. 152), changes in status ([9], pp. 1–19), and ritualistic intervention ([1], p. 203) are examples.

Extrusion may also occur in a variety of ways. The execution or permanent ostracism of an offender of the social mores from the community are not uncommon modes for dealing with serious forms of deviancy. Prisons and mental hospitals in this society provide a means for extruding a deviant individual from a community. A deviant individual may be removed to such a setting on either a temporary or permanent basis.

It seems probable that any given social-cultural system has tolerance levels that, if exceeded, result in decisions to extrude the offending individual from community life. These tolerance levels are probably variable not only from one society to another but within any given society. The tolerance levels and any decisions to extrude an individual must be explained within the social system, and such explanations must be in accord with the prevailing value assumptions of the society. It would, for example, be incongruous in this society to imprison a delusional schizophrenic on the sole basis that he was disruptive. There is a social value which holds that since he is mentally ill he is not responsible for his behavior and cannot be held accountable. At the same time, it seems perfectly rational to remove him for his own good in that he cannot, in his delusional state, make a rational decision to seek treatment. Such a rationale for extrusion would be unlikely if this society viewed the behavior of schizophrenics as willful acts predicated upon a conscious and premeditated rejection of reality.

Not all mentally ill individuals are involuntarily committed to state mental hospitals. Extrusion is selective. Edwin Lemert ([7], p. 67) suggests that on the basis of the available literature only 50% of 'mentally diseased' persons are ever institutionalized. He further suggests that:

> The tolerance of groups for the disruptive behavior of psychotic members finds expression in variable institutionalizing tendencies or preferences, or, perhaps better, in a tolerance quotient ([7], p. 68).

It is now possible to advance a tentative hypothesis to explain the functional relationship between the mental health codes and the criminal justice system, and this can be done without considering the second difference between the two conventions. Mental health codes may be viewed as social instruments for extruding deviant individuals whose behavior exceeds as yet undefined tolerance levels in the social system. Mental health codes and the criminal justice system may be complimentary in that they address different and distinct populations, providing parallel mechanisms for extruding disruptive individuals from the community. The redundancy of the two systems can be explained on the basis of the meaning of the behavior attributed to the two populations as that behavior is understood in relation to the value assumptions underlying the social structure: the insane are ill and not accountable while the criminal are willful transgressors. This hypothesis will be sufficient only if it is possible to show that the populations with which the mental health codes and criminal justice systems deal are in fact distinct. There is reason to doubt that such is the case.

There has been a growing tendency within this society to redefine certain deviant behaviors as symptoms of mental illness rather than as criminal acts. So called sexual perversions, alcoholism and juvenile delinquency serve as examples of behaviors now considered as evidence of mental illness, but these were previously viewed as willfully criminal. Although this re-definition may reflect an increasing sophistication of psychiatric diagnostic ability, such a conclusion seems premature. It at least suggests that the boundaries between populations served by the two conventions may shift with changes in the social system itself.

A conclusion that the two conventions are directed at distinct and separate populations becomes even more problematic when the net effect of transactions between representatives of the two conventions are considered. The author of this essay has been intimately involved in the proceedings of both mental health codes and the criminal justice system. He has observed that while some portion of the population processed under the two conventions appears distinct, there is a large area of overlap in which circumstances determine which convention will be invoked to deal with a given individual.

District attorneys, judges, police, lawyers, psychiatrists and representatives of social service agencies compose and maintain an informal communication network in most communities which is active in determining which convention will process or deal with allegedly deviant individuals. It is not uncommon for representatives of mental health service systems to abandon an individual previously diagnosed as mentally ill and in essence turn him over to the courts for involuntary commitment or trial in the criminal justice system. Negotiations are carried out between the district attorney and mental health representatives to drop criminal charges to clear the way for civil commitment. Frequently judges, lawyers, and psychiatrists, in order to save time, turn civil commitment proceedings into pro forma rituals.

The second difference between the two conventions will be of assistance in conceptualizing a functional relationship between the two conventions that does not depend on processing distinct populations. The two conventions differ in the judicial rules that must be followed in reaching a decision to remove an individual from the community. The specific differences have been discussed elsewhere in this volume and no attempt will be made here to provide an exhaustive description of the different rules of evidence and means of proceeding in the two conventions. Suffice it to say that the criminal justice system provides extensive safeguards for the alleged offender and operates in such a way as to free the guilty rather than punish the innocent, while mental health codes provide much less in the way of legal safeguards for the alleged insane and operate in such a way as to hospitalize the non-dangerous rather than free the deranged. This difference is overtly explained on the basis of the state's motivation, i.e., to punish the criminal and help the ill.

The greater subjectivity in making a decision that is provided in the various mental health codes and the relative lack of legal safeguards provide a much more flexible approach for removing deviant individuals from community life than is possible under the criminal justice system. There is little data available for making a judgment as to whether the flexibility inherent in the various mental health codes plays a part in the ongoing social re-definition of criminal behaviors as sick behaviors. It should be noted, however, that in criminal proceedings in which an individual is excused from responsibility for a crime on the basis of insanity, subsequent incarceration is indeterminant and totally dependent upon the subjective judgement of his physician, who is himself subject to pressures from judges, law enforcement officials, representatives from various community institutions, family members and ultimately the populace as a whole. In this regard Ronald Leifer ([6], P. 140) has noted that one function of the state mental hospital is "... to confine, control, supervise, and reform deviant individuals as a device for circumventing the restraint that Rule of Law places on the power of the state". He goes on to say:

> A second function of the mental hospital is therefore to protect our image as a free society, devoted to Rule of Law. It permits us to hide violations of the principle of individual freedom under the law in order to improve our internal security and tranquility. The image of our nation as a free society would be badly tarnished if it were openly recognized that an individual could be confined without trial, bail, or due process by means of psychiatric commitment ([6], p. 140).

Whether or not re-definition of criminal behavior results in whole or in part from a desire to deal with deviant behavior in a more flexible manner than is possible under the criminal justice system, the result is the same. Once re-defined as evidence of illness, such behaviors and the individuals who manifest them can be denied the rigid legal safeguards of the criminal justice system while they are being extruded from community life.

It would be simple at this point to conveniently indict the various mental health codes and the modes in which they are exercised as social abuses, to see in them a Machiavellian attempt to undermine national principles of individual rights under the law. Many observers have reached just such a conclusion. The question remains, however, why the duplication evident in the two conventions exists. It would seem more efficient and less cumbersome to amend the criminal justice system in some manner that would provide less in the way of legal safeguards for the accused and provide for more subjectivity in making decisions as to whether a given individual should be extruded. There are, in fact, advocates of just such an amended criminal justice system, calls that have not been able to achieve any substantial reduction in the rights accorded an accused criminal.

Before attempting a final, functional answer to the question of the duplication in the two conventions, a brief review of the arguments presented may be helpful. All societies must deal with deviancy, and extrusion of deviant individuals is one means of dealing with serious deviant behavior. Prisons and state mental hospitals provide a means in this society for temporary or permanent extrusion of deviant individuals. Mental health codes and the criminal justice system define alternative sets of rules for reaching a decision to extrude any given individual. When extrusion occurs a tolerance level in a given community has been exceeded. The question here is not what communities should tolerate in the way of behavior but rather how community tolerance levels can be dealt with in such a way as not to compromise the egalitarian values on which this society is founded.

Extrusion of deviant individuals must be rendered consistent with existing social values, those basic value assumptions that define and give meaning to the social structure itself. If they are not, then the social structure is at risk and either the available mechanisms for extrusion or the underlying social values or both must change. Values and social mechanisms exist in some sort of equilibrium. When, as is inevitable in a dynamic social system, there are stresses between social mechanisms and social values, those stresses will create irresistible forces in the direction of restoring equilibrium ([3], p. 634).

The mental health codes of the various states provide a means of by-passing the rigorous safeguards of the criminal justice system which is firmly based upon egalitarian principles and the rights of individuals. These codes do so, however, on the basis that some individuals are ill and not simply bad. They do so by defining the population they relate to as distinct and separate from that dealt with in the criminal justice system. By a simple manipulation of definition they provide protection for the criminal justice system from the recurrent popular demands for modification of the safeguards of individual rights that are built into that system.

If the various mental health codes were abolished, as some observers have suggested, what would be the result? Presumably communities would not simply increase their tolerance for deviant behavior. One might assume that something of the kind would occur, but that there would also be increasing pressure on various legislatures to expand the scope of behavior defined as criminal and on the judicial system to ameliorate and modify procedures that provide safeguards for the accused. In a very real sense mental health codes may be viewed as an escape valve that makes possible a society's self-righteous, albeit hypocritical, regard for individual rights.

The importance of such a conclusion is twofold. First, any unwitting and impulsive reform of mental health codes may have unexpected and undesired consequences for the criminal justice system. Secondly, without careful monitoring, mental health codes may ultimately undermine the rights incorporated in the criminal justice system by providing a too convenient alternative. Any future efforts at reform should consider the two conventions in relation rather than as distinct entities with a similar aim of extruding individuals with deviant behavior from the daily social transactions of a community.

University of Texas Medical Branch
Galveston, Texas

BIBLIOGRAPHY

1. Clark, M.: 1970, *Health in the Mexican-American Culture*, University of California Press, Berkeley, California.
2. Cohen, A.: 1966, *Deviance and Control*, Prentice-Hall, Englewood Cliffs, New Jersey.
3. Davis, K.: 1948, *Human Society*, MacMillan, New York.
4. Evans-Pritchard, E. E.: 1940, *The Nuer*, Clarendon Press, Oxford, England.
5. Goffman, E.: 1961, *Asylums*, Aldine, Chicago, Illinois.
6. Leifer, R.: 1969, *In the Name of Mental Health: The Social Functions of*

Psychiatry, Science House, New York.
7. Lemert, E.: 1967, *Human Deviance, Social Problems and Social Control*, Prentice-Hall, Englewood Cliffs, New Jersey.
8. Merton, R.: 1957, *Social Theory and Social Structure*, Free Press, New York.
9. Newman, P. L.: 1964, ' "Wild Man" Behavior in a New Guinea Highlands Community', *American Anthropologist* **66**, 1–19.
10. Rosenhan, D.: 1973, 'On Being Sane in Insane Places', *Science* **179**, 250–258.
11. White, L.: 1975, *The Concept of Cultural Systems*, Columbia University Press, New York.
12. Wing, J. K.: 1967, 'Institutionalism in Mental Hospitals', in J. Scheff (ed.), *Mental Illness and Social Processes*, Harper and Row, New York.

STEPHEN WEAR

THE DIMINISHED MORAL STATUS OF THE MENTALLY ILL[1]

Throughout this volume, a subtle but basic conception of the moral status of the mentally ill is operative; i.e., that the special characteristics of mental illness involve a diminished status as moral agents for those thus afflicted. Recalling that the notion of moral agency generally assumes certain levels of rationality and freedom, or cognitive and volitional competence, we may note diminished levels of such competencies being presupposed by authors in this volume as they reflect on the situation of being mentally ill. Moore, for example, wants to speak of 'degrees of irresponsibility' and of a severely diminished responsibility in the mentally ill ([16], p. 25), Neu, for his part, speaks of 'diminished responsibility' at law for the mentally ill ([17], p. 77), that there are 'degrees of freedom' ([17], p. 89), and that a notion of 'partial insanity' may be useful and appropriate for dealing with the mentally ill in the legal setting ([17], p. 92). And, in a similar vein, Brock believes that competence is a matter of degree ([2], p. 170), and that the mentally ill have "an impaired capacity to be guided by legal rules" ([2], p. 156).

Three initial points should be made regarding this talk of diminishment. First, one should note the import of the language being used here. In most naive accounts of moral agency, a binary approach is taken; i.e., either one is a moral agent, invested with special rights and responsibilities, or one is not (and one loses such rights and duties). But here we see talk of degrees of rationality, freedom, agency, and competence, of impaired capacities, and diminished responsibilities. Such language signifies a graded view of moral agency displayed over a continuum on which a non-reflective, unfree beast occupies the lower limit, and a normally rational, self-determining person occupies the upper limit of full moral agency. Individuals suffering from cognitive and volitional incapacities are seen to fall somewhere in between. Second, it must be regretted, however, that the contributors to this volume have not given any account or argument for such a notion of diminished moral status (or even paused to explain what is meant by the preceding language). But, third, it should be noted that such language does, in fact, mirror popular views on the mentally ill, as well as legal and institutional practice.[2] I will expand the last point first; the former points concerning the need for a systematic account and justification of the notion of

B. A. Brody and H. Tristram Engelhardt, Jr. (eds.), Mental Illness: Law and Public Policy. 221–230.

diminished moral status can then be better appreciated and reflected upon.

In the sphere of institutional practice, the mentally ill are clearly treated as if their moral status were diminished. They are confined and forced into certain regimens (chemical and social) without their consent and often against their will. Moreover, they are often excused from blame and liability for their actions, both before and during commitment. The rights and duties of the 'normal' person are not applied to them. Further, such institutional responses to the 'mentally ill' seem to be generally accepted. Apart from the problems of where lines are to be drawn, or the potentials for abuse (political and otherwise), the consensus regarding the mentally ill is that they are defective both in rationality and will, that they are dysfunctional human beings.[3] It is generally accepted, therefore, that confinement and treatment are necessary both for their own sake and that of others. At the same time, certain criminal and civil proceedings against the mentally ill are seen as inappropriate; generally, they are not held responsible for the injuries which they inflict upon others.

A further look, especially at institutional practice, reveals an even more complex response to the mentally ill. First, even the most hostile, incoherent patient still retains rights to food, exercise, proper sanitation, etc. though he may be locked in a single room and heavily sedated. Though the privileges and duties of the mentally ill are diminished, they are never totally lost. Secondly, not only do all patients, even the most hostile or incapacitated, retain certain basic privileges, but the full context of privileges and obligations varies from individual to individual. From the heavily sedated patient in isolation to the patient set at liberty in an outpatient situation, the privileges and duties of individual patients form a hierarchy. For example, most patients are allowed to move around their own unit; some have the liberty of the hospital grounds; and some are free, for varying periods, to leave the hospital. Equally, a patient's duties are similarly variable as well as correlated with his or her privileges. By fulfilling dormitory and program responsibilities, which also vary, the patient retains and may increase his privileges, while failure to do so or inappropriate behavior can lead to the reduction or loss of such rights.

The history of judicial findings regarding the mentally ill and the retarded reflects a movement towards a more nuanced appreciation of the grades of mental incapacity. Though there is no uniform and unambiguous trend, one sees an attempt to recognize the mentally incapacitated as retaining some of the rights and privileges of normal persons.[4] There has, in fact, in some cases, been an attempt to have the loss of rights and duties be strictly commensurate with the state of the individual involved. One might refer here to a recent court ruling that mental retardation was not, in and of itself, a bar against the privileges of franchise. The right of retarded adults to vote was upheld, and all that was required was that they satisfactorily complete voter registration forms (*Carroll v. Cobb*, 35 A. 2d 355 (N.J. Super. Ct., App. Div., Feb. 23, 1972)). Other cases, though not as unambiguous, show as well a move toward giving special consideration to the degree of incapacity and the sort of right involved. In Texas, a retarded woman who was promiscuous and had had several children as a result, sought a court order for sterilization on the urgings of other concerned parties. The court held that she was not competent enough to make such a request, though it was pointed out, in further deliberations, that she was competent enough to raise her children and thus retain her parental rights over them.[5] In Pennsylvania a schizophrenic patient refused to consent to surgery for carcinoma of the breast. This right to refuse treatment was upheld because the patient was determined to be competent when she first made the refusal, though she subsequently became delusional (*In re Appointment of a Guardian of the Person of Maida Yetter*, Docket No. 1973–533 (Pa. Ct. of Common Pleas, Northampton Co. Orphan's Ct., June 6, 1973)). Conversely, a person committed as a criminal sexual psychopath gave consent to innovative psychosurgery, but it was held that this consent was not valid, because the patient's capacity to consent was diminished by the long-term institutionalization he had suffered (*Kaimovitz v. Department of Mental Health for the State of Michigan*, Civil Action No. 73–19434–AW (Mich., Wayne Co. Cir. Ct., July 10, 1973)). Though all these people had diminished rights and responsibilities by virtue of their mental incapacities, it was nonetheless determined that they retained certain rights in other areas. Their abstract status as mentally ill or deficient was not a sufficient determinant; it became a question of the level of competence and the kind of right involved. In the case of the Pennsylvania woman, for example, there was no question that she could not be set at liberty, but she was judged sufficiently competent to refuse the surgery.[6]

The notion of a graded, non-binary account of moral agency, affording a range of varying diminished moral status to the mentally ill, is thus suggested by such determinations in the courts, by institutional practice and popular views, and, given the context of this volume, by the unargued use of such a notion by many of its contributors. But is such a notion valid or even coherent? What questions and problems arise as one attempts to construct such an analysis? The remainder of this paper sketches out the direction in which we must go in order to formulate such a notion and suggests some of the constraints and difficulties which threaten such a program.

In order to flesh out such a view of graded moral agency, one must first set out the lineaments of full moral agency – its elements, and the grounds that make it more than merely stipulative. First, basic to the ethical situation is the notion of responsibility. Moral agents are responsible for their actions; praise and blame, in a moral sense, are applied to them. Indeed, if responsibility is to be meaningful, one must consider people to be free. It makes no sense to praise or blame a person for what one could not help doing nor, for that matter, for not doing what one could not have done. In other words, one must be capable of freely choosing between alternative courses of action; without alternatives, or free choice, one can hardly be held responsible. Also implied by such an analysis is a corresponding capacity for rational deliberation; a moral agent must be capable of understanding the nature of his or her situation, the actions[7] he or she performs and, to some extent, the consequences of those actions. Without such an understanding, responsibility is not attributable; if one does not know what one is doing, either factually, or morally, then one is not responsible. In sum, the status of a moral agent is appropriate only to a being who can freely and rationally choose a course of action; freedom and rationality are thus necessary conditions for the moral life.[8]

The preceding is simply analytic to the notions of responsibility and moral agency. If these notions are to be meaningful, in theory and in application, then freedom and rationality are their necessary conditions. Such an analysis, of course, does not prove the existence of the moral life; rather, it only asserts that if there is such life, then freedom and rationality are required capacities for its participants. It might be noted here, however, that a determinist would certainly deny such freedom, and one who is especially impressed with the role of unconscious factors in determining behavior would severely depreciate the effectiveness of rationality. But the choice here is not an innocent theoretical one. By denying freedom and rationality, one removes the sense of most of the law and morality, as the preceding analysis is meant to imply. In the absence of sufficient proof to the contrary, then, I would recommend that freedom and rationality be received as ontological principles. Further, it should be emphasized that I do not intend that the capacities for freedom and rationality exhaust either the meaning of personhood or that of moral agency. Surely there is more that we should have regard for in persons, and the ethical status and situation of the individual is more complex. But I do offer these two elements as constituting the core sense of both notions.

Moving to a consideration of mental illness itself, the first area of concern is conceptual. That is, before we attempt to pass on to the moral status of the mentally ill, we should define mental illness. But, of course, this volume testifies to the fact that there is an even more immediate question at hand. Is there any such thing as mental illness at all? Throughout this volume and contemporary psychology, one can sense the figure of Thomas Szasz lurking in the background. A number of authors in this volume have exercised themselves in countering Szasz's claim that mental illness is a myth and, by implication, that the 'mentally ill' should be treated no differently from other people. Corinna Delkeskamp has argued that these other authors have not sufficiently demolished Szasz's position ([10], p. 181). I refer the reader back to the text on this issue. I would, however, like to make one observation of my own concerning Szasz's position.

One of the *prima facie* conclusions regarding many of those classified and incarcerated as mentally ill is that they are severely incapacitated human beings. Now, if this is true, if we have people who are in fact severely irrational, governed by bizarre, irresistible impulses, then I cannot see but that some moral decision must be made to excuse them from blame and liability for their actions, and also, at least in some cases, to act paternalistically for their involuntary commitment. Szasz, of course, wants to say that the concept of mental illness is not a descriptive term, but is prescriptive ([20], p. 235). In other words, when we say someone is mentally ill, we are not saying he is sick as in physiological illness; we are naming a role ([20], p.235) and justifying a societal response to it. Thus, mental illness is a social construct utilized by the majority to institutionalize and degrade as sick those who are only different.

It is surely not possible to offer a full response to such a view here. It should be enough to raise the following question: beyond the political and societal needs served by such policies, can we not still say that many such people are, in fact, sick? The impression of incapacitation, bewilderment, fear and lack of patterned behavior witnessed in many mental patients is so obvious and spontaneous that I have serious doubts about the validity of Szasz's arguments. Further, even if we were to agree with R. D. Laing, for the purposes of argument, that mental illness is an adaptive response to an insane environment, the response is no less pathological for being such an adaptation. It is one thing to point out the more subtle causes of mental illness, and the intrusion of our own arbitrary societal norms in its designation; it is quite another thing to infer that none of these people are sick or dysfunctional as they so obviously seem to be. In this regard, Baruch Brody has put the point elegantly in the previous volume of this series; i.e., that though Szasz may be right, he has yet to prove it ([3], p.251).

It should be noted as well that Szasz views the choice between 'mental illness' and 'problems in living' as exclusive – *tertium non datur*. The spirit of the analysis of this essay would urge a more nuanced view, i.e., that many mental patients 'suffer' partially from problems of living *and* partially from a disease state, and thus one would see a view of partial responsibility suggested again by the incapacitating presence of true mental disease. Szasz's insights into the political and cultural manipulation of mental illness designations are appropriate and merit more serious reflection, as Delkeskamp suggests ([10], p. 200, n. 1). However, a more 'realistic' account of the situation would acknowledge that what has often been called mental illness represents modes of political control or at least a failure to adapt to one's culture (i.e., problems in living), while acknowledging that much of what is termed mental illness is still genuinely pathological. These terms are not exclusive in the case of particular individuals. An individual termed 'mentally ill' may, in fact, have both a 'problem in living' and a true pathology. Thus, one can maintain Szasz's insight (see Creson [8]), that much of mental illness talk represents modes of social control, without saying that it is all social control. In short, Szasz's insights can be maintained if one talks of grades of problems in living and grades of mental illness. It is quite appropriate to consider that many cases of 'mental illness' have significant adaptive or social and political factors underlying them. But Szasz has overreacted to such factors, and thus errs in not allowing us to recognize and deal with the genuine pathological conditions which, however 'impure', are simply present in some cases.

How then should one proceed with the explication of mental illness, especially in terms of the previous analysis of moral agency? It has been noted that freedom and rationality are necessary conditions for moral agency and responsibility. It has also been suggested, in line with traditional views, that the freedom and rationality of the mentally ill are severely diminished. And, since mental illness is being seen as a deviation from the norm, the next step is to see if one can formulate a concept of what is normal.

Joseph Margolis's paper, 'The Concept of Mental Disease: A Philosophical Examination' [15], is instructive for this further task of determining the norms of mental health.[9] Is there, as Margolis wants to maintain, any "normal or characteristic or favored forms of life?" ([15], p. 7). We find such a notion implicit in the law in the standard of the 'reasonable man', as noted in the discussions of Neu ([17], p. 76) and Coleman ([7], p. 111). Different accounts of the background assumptions of the construct 'reasonable man' could be offered. In this volume, Margolis proffers one implicitly in his talk of prudential values as those which are normative for human nature ([15], p. 20). Margolis, while speaking of these prudential values, talks critically of species typical facts as a basis for a normative view of human function ([15], p. 20).

But, here again, a qualified view is in order. One must acknowledge the variability of such notions and the resulting variations in the application of such notions toward determining the status of a given individual. Human function, in our post-Darwinian world, may be best construed in terms of adaptation to particular environments, and since there is no standard environment, there will be no standard of adaptation. Also, there will not even be a standard description of proper function, since different performances will have different results in different environments. Moreover, if one is concerned with species survival, one would hope that there are some genes for phenotypes not perfectly adapted to current environments so they can function as security for adaptation and thus survival in possible future environments. One would expect grades of 'normal' mental function and adaptation; having an obsessive-compulsive personality type, for example, will, in varying degrees, count as health or disease in varying environments. Prudential values of survival of the individual or of the species will allow one to identify success, but success, in concrete terms, will have different meanings and nuances in different cultural and environmental milieus [12].

Reviewing the argument of this essay, it was first pointed out that institutional and legal practice treat the mentally ill as having a diminished moral status. It was further noted that a more nuanced view of such a situation was appropriate, i.e., that an all-or-nothing approach to moral agency and mental illness should be avoided in preference for a view of them as graded continua. Freedom and rationality were then seen as the essential notions by which moral status and mental illness could be interrelated. In effect, the suggestion was that there are grades of freedom and rationality, and of rights and responsibilities, and they may be best seen as varying in direct proportion to each other. Further, Szasz's view that mental illness is a myth, a social construct without descriptive content, was strongly qualified. Beyond problems of living, as he put it, it was maintained that it was still appropriate to talk of a genuine pathological core as extant in many mentally ill persons. Finally, a more discriminating view of a normative conception of human nature was outlined. It was seen that no strict standards of prudential values, proper human functioning, or adaptive 'success' could stand, given the evolutionary situation and the lack of a standard environment. But more nuanced normative concepts may be formulated, both on the most abstract level of prudential values or, more concretely, within a particular cultural and environmental matrix. Given such variable norms, the final suggestion of this paper is, then, that they might be related to the individual mental patient, and that an appropriate determination of his diminished moral status might be displayed in terms of the degree to which he fell short of such norms. To the extent that the freedom and rationality of a mentally ill person are diminished, to such an extent are his rights and responsibilities degraded.

In conclusion, it is held that such more sophisticated and nuanced conceptions of moral agency and mental illness are required, that they tend to mirror institutional and legal practice, and reflect popular views on the special characteristics of mental illness and the moral implications involved. Politically, such an account would also facilitate the growing concern and efforts on behalf of the mentally ill. For example, Szasz's suggestion, as Delkeskamp puts it, to render people responsible by treating them as if they were responsible ([10], p. 198), could be respected, but again only to the extent that the patient retains sufficient competence for a given responsibility; it will certainly not help if *our* expectations for some patients become delusional.

Just as the competent therapist, in constructing a patient's program, takes into account the individual characteristics of that person's affliction, so may the individual patient's rights and responsibilities be tailored to his level of competence. The preceding surely complicates the situation, and it would be nice to have strict guidelines for pigeonholing people, but the facts of the matter require something quite different from this.

State University of New York at Buffalo
Buffalo, New York

NOTES

[1] I want to thank Edmund Erde and James Speer for their substantial aid in generating this article. Special thanks are due, further, to H. Tristram Engelhardt, Jr. for our many arguments concerning this topic, which helped forge this paper.

[2] Moore's article [16] is especially helpful here, both in rehearsing the history of legal thought on mental illness, and in noting the role of popular sentiments in this debate.

[3] Important work has been done concerning the causal relationship of popular views and institutional practice regarding the mentally ill and an appreciation of this interaction can not help but sophisticate the discussions in this area. See especially [4, 6, 9, 11, 13, 18, and 19].

[4] Although no work clearly details this development in legal thought, several of the

basic historical studies, cited above in note 3, also reference and suggest these trends. See especially Deutsch ([11], pp. 332–441) and Caplan ([6], pp. 117–125 and pp. 233–246).

5 *Frazier v. Levi*, 440 S. W. 2d 393 (Tex. Ct. Div. App. 1969). See, for a general discussion, Charles H. Baron ([1], pp. 267–84, especially p. 274).

6 I do not intend to signify that these cases represent any specific trend in the law, especially as regards the concern to correlate particular rights and responsibilities with levels of competence. This has not yet been researched and documented.

7 See Brock's paper [2] for extremely helpful reflections on these matters.

8 This sort of analysis is, of course, a traditional one in philosophy, and its classic exposition was laid out by Kant. I have avoided citing Kant here, because I do not want to make an excursion into a complex, scholarly area when, I believe, such an analysis can be simply offered and received as valid, at least for our purposes here. One might want to consult especially the first chapter of Kant's *Foundations of the Metaphysics of Morals* for further reflection on this sort of argument.

9 Margolis's view of the moral implications of his "normative findings" is, however, not made clear here. Also, I do not mean to suggest that he would agree to the use which I propose for such findings. The reader is strongly encouraged to refer to Margolis's article [14] for further discussion of his view. See especially his remarks on "the functional norms of psychiatry" on p. 245.

BIBLIOGRAPHY

1. Baron, C. H.: 1976, 'Voluntary Sterilization of the Mentally Retarded', in A. Milunsky and G. J. Annas (eds.), *Genetics and the Law*, Plenum Press, New York, pp. 267–285.
2. Brock, D. W.: 1980, 'Involuntary Civil Commitment: The Moral Issues', this volume, pp. 147–173.
3. Brody, B.: 1978, 'Szasz on Mental Illness', in H. T. Engelhardt, Jr. and S. F. Spicker (eds.), *Mental Health: Philosophical Perspectives*, D. Reidel Publ. Co., Dordrecht, Holland, pp. 251–261.
4. Burnham, J. C.: 1960, 'Psychiatry, Psychology, and the Progressive Movement', *American Quarterly* **12**, 457–465.
5. Burnham, J. C.: 1967, *Psychoanalysis and American Medicine, 1894–1918: Medicine, Science and Culture*, International Universities Press, New York.
6. Caplan, R. B.: 1969, *Psychiatry and the Community in Nineteenth-Century America*, Basic Books, New York.
7. Coleman, J. L.: 1980, 'Mental Abnormality, Personal Responsibility and Tort Liability', this volume, pp. 107–133.
8. Creson, D. L.: 1980, 'Function of Mental Health Codes in Relation to the Criminal Justice System', this volume, pp. 211–219.
9. Dain, N.: 1964, *Concepts of Insanity in the United States, 1789–1965*, Rutgers University Press, New Brunswick, New Jersey.
10. Delkeskamp, C.: 1980, 'Critical Use of Utilitarian Arguments: Szasz on Paternalism', this volume, pp. 177–207.
11. Deutsch, A.: 1949, *The Mentally Ill in America*, Columbia University Press, New York.

12. Engelhardt, H. T., Jr.: 1976, 'Ideology and Etiology', *Journal of Medicine and Philosophy* 1, 256–268.
13. Grob, G. N.: 1965, *The State and the Mentally Ill: A History of Worcester State Hospital in Massachusetts, 1830–1920*, University of North Carolina Press, Chapel Hill, North Carolina.
14. Margolis, J.: 1976, 'The Concept of Disease', *The Journal of Medicine and Philosophy* 1, 238–255.
15. Margolis, J.: 1980, 'The Concept of Mental Illness: A Philosophical Examination', this volume, pp. 3–23.
16. Moore, M. S.: 1980, 'Legal Conceptions of Mental Illness', this volume, pp. 25–69.
17. Neu, J.: 1980, 'Minds on Trial', this volume, pp. 73–105.
18. Rosen, G.: 1968, *Madness in Society*, Routledge and Kegan Paul, London.
19. Rothman, D. J.: 1971, *The Discovery of the Asylum: Social Order and Disorder*, Little, Brown and Co., Boston, Mass.
20. Szasz, T.: 1978, 'The Concept of Mental Illness: Explanation or Justification', in H. T. Engelhardt, Jr. and S. F. Spicker (eds.), *Mental Health: Philosophical Perspectives*, D. Reidel Publ. Co., Dordrecht, Holland, pp. 235–251.

JAMES SPEER

A CONCERN FOR HARDENING OF THE CATEGORIES[1]

By keying philosophy to the designs of law and public policy the editors of this volume align it squarely with the interests of many legislators, judges, district attorneys, police, lawyers, psychiatrists, and workers in social service agencies. The most important and urgent issues of mental illness that concern these varied groups all have to do with conceptual, moral and ethical questions peculiarly within the special competence of philosophers. Studies of these linkages by philosophers, and other scholars dealing with essentially philosophical concerns, therefore, could be informative and perhaps even influential. Yet if ever these hopeful, exciting prospects are to be realized, much more will have to be said about matters of perspective and proportion, of context and structure. The virtual silence of the contributors to this volume on these scores indicates a general problem: the legal and public policy aspects and ramifications of mental illness have never been fully surveyed, mapped and catalogued. Along the artificial lines of categories within various disciplines, to be sure, tiny plots and subdivisions have been staked out. But from these alone one cannot determine if the results are gerrymandered or simply crazy-quilted. With so little appreciation of the general issues of mental illness, it is exceedingly difficult to appraise the accuracy and relative significance of studies on particular issues. Questions about the relationships among parts, and of parts to whole, desperately need systematic, sustained attention.

Lack of medical and scientific consensus regarding the nature, causes, and mechanisms of mental disorder obviously complicates, but has not created, these difficulties in geographic puzzle-solving. Certainly there are matters about which we have some clear knowledge, and many others about which we have footnoted opinion. Library shelves sag with the weight of books and journal articles devoted to the subject of mental illness and to the conceptual, moral, legal, social, and political problems that stem from our treatment of the mentally ill. Yet for all this heavy and ongoing investment of pulp and ink, we have little sense of context and framework. Slow progress in these respects would be discouraging enough, even if the effects were confined within the families of humanities and social science disciplines. It is quite a more serious thing that our uncertainties about structure, connections and

B. A. Brody and H. Tristram Engelhardt, Jr. (eds.), Mental Illness: Law and Public Policy. 231–237.

interrelationships bedevil those who make, interpret, and enforce public policy.

Commendably, the editors are dedicated to reaching and assisting these groups. This volume's essays, deriving as they have from symposium papers for mostly non-academic audiences, attempt effective communication. Greater success in future efforts, I believe, will depend upon our formulation of an analytical context broad enough to engage the full range of issues, alternatives, constraints, and possibilities that exist in the realms of law and public policy. Merely semantical debates or purely negative criticisms of present practices or procedures, as Daniel Creson [2] points out, help very little.[2] Needed, instead, are generalized frameworks of inquiry. Comprehensive structures for classification and analysis would facilitate an understanding of the relationships among issues and needs, not only as they lie within the preserves of various disciplines, but also as they fall under the authority of particular departments, agencies or branches of the law. I do not suggest that such approaches could be developed quickly or easily made explicit. But there is a need for a broader perspective. My comment is focused on this need in the area of law.

Essays in this volume by Professors Moore [4], Neu [5], and Coleman [1] deal with particular conceptual distinctions that bear on the project of formulating the integrated, comprehensive view I have urged. In terms of each essay's subject matter, the policy implications of these concepts are weighed with some care. Too little thought, however, is given to the reach of these implications beyond the particular branches of law under study. It is unfortunate that they did not seize the opportunity to generalize beyond their immediate, particular foci.

The introductory section of Moore's paper, which outlines many of the categories in which mental illness comes before the courts, reminds us that we do not have a general, integrated concept of mental illness across these categories. As he explains, while the defense of insanity in the criminal law and questions of tort liability and involuntary commitment in the civil law clearly involve the most conspicuous resorts to mental illness, the concept itself is expressly utilized in other branches of the law. Moreover, courts in many other substantive areas deal with questions of the degrees of competence and capacity without consideration of an individual's abstract status of being mentally ill or not. The criteria for decision and likely outcome of any given case involving the same person's competence and capacity hence may vary widely from one branch of law to another.

Stephen Wear has observed one aspect of the broader analytical context by referencing actual cases that illustrate a graded continuum of retained rights and responsibilities by the mentally impaired ([7], pp. 222–223). Both with respect to Moore's outline and Wear's perceptive analysis, consider the following hypothetical situations which starkly illustrate the possibilities created in absence of this categorial integration. Consider, given several different sets of circumstances, what might happen to the same individual assuming only an unchanging mental state and consistent behavior: certain of this person's contracts are declared void. He is successful, however, in resisting the attempts of relatives to gain a court-appointed conservator, and is permitted to continue managing his property. Claiming himself to be mentally ill, and seeking to have his insurance carrier pay the costs of treatment of his 'condition' as an 'illness' covered by the policy, he wins the lawsuit. Yet, in civil commitment proceedings, he claims just as strongly *not* to be mentally ill. He is not deprived of his liberty. Another person, though, is given proxy authority to consent to lifesaving medical treatment he would not approve. Divorced some years before, he is determined to lack sufficient capacity to enter a valid marriage. He is permitted to retain, however, custody of and parental rights over his children by the former marriage. All these findings, in another sequence or in almost any other combination of outcomes, might arise before, during or after his being found 'legally insane' in criminal proceedings or liable for damages in a tort action. Validity of his will, depending naturally on the time and circumstances of its execution, might or might not be determined with reference to any of the foregoing.

The difficulty in Wear's actual cases and in my highly improbable but nonetheless possible ones, is not that purposes served by one area of law may, and often do, differ greatly from policy aims in another branch of the same state's legal system. Rather, the point is that the potential for such widely varying results regarding the same individual enormously complicates the business of formulating policies sensitive to the needs of *classes* of persons. The law's various categories, forms of action and procedures peculiar to them are seen to serve legitimate and justifiiable ends. On this basis, such branches or categories of the law are understandably considered and studied as being distinct from one another. Still, the law's substantive divisions do not create airtight compartments. To treat them as such is to risk being disabled by a hardening of the categories.

By looking only at a single area, it is difficult to see the point at which determinations proper for one branch of law overlap, confuse, or undercut the equally (if not more) important social policies served by another branch of law. An analytical approach that transcends the law's several categories, I believe, would encourage review of inconsistent or competing social aims. Such an approach would open consideration of whether the criteria and logic of decisions might be harmonized. Moreover, such an approach would aid in judgments on whether findings in one branch of law ought to be noted and given weight in another. With attention to people as well as fields of law, we might define with greater confidence the exclusivity of group labels for mentally ill, mentally defective, mentally retarded, and the like.

I am reinforced by Neu's argument in my conviction of the necessity for a broader analytical perspective: "The main point to take hold of is that . . . of social policy, it is a mistake to consider questions of criminal confinement separately from issues of civil commitment and other forms of social control. The underlying problems and the institutional responses are tied" ([5], p. 93). I would elevate this point to a rationale for sustained inquiry across still other substantive boundaries. Now in many of these areas, of course, the differences are great and the boundaries are sharply defined. As Coleman emphasizes, the requirements of justice in torts differ fundamentally from those in contracts and the criminal law ([1], pp. 121–126). Even so, we make these distinctions largely on the basis of social policy and particular characteristics of the individual's claim of rights and remedies. Before troubling to look, I should think it premature to despair of ever finding opportunities for linkages and generalizable terms, concepts and criteria. Moore cautions that the law's criteria are for the law to settle, and that the law must define legal concepts for itself ([4], p. 36). Certainly one would not expect uniformity for all areas of law, or some kind of a separate system of jurisprudence for persons of less than full mental competence or capacity. Nor, as Moore points out, can the terms and concepts of social scientists be safely adopted and plugged into any legal formula ([4], p. 36). Remaining, however, is the question of whether *from within the law itself* such integrations of concepts, terms, and procedures might not be advantageous.

We shall not have an answer about potential benefits, of course, until the cross-categorial integrations are fully studied and developed as theoretical matters and, eventually, refined and implemented as policies. Whether such products of work from a broader analytical perspective would improve and sophisticate the law's handling of persons of diminished faculties is a question about which one can only speculate. My concern for a hardening of the categories in these essays is not that the basic question is left with a tentative answer, but that it was never asked. Social purposes and the full range of an individual's legally significant activity and states of mind are not accommodated by any single one of the law's substantive divisions. Preoccupation with one branch of the legal system to the exclusion of all others will enhance neither our understanding nor prospects for constructive dialogue with policy-makers and enforcers.

Some persons will likely disagree. Others might raise a challenge on antecedent grounds – why bother? Even if different branches of the law evaluate and treat the same person in highly inconsistent and perhaps grossly unfair ways, are these cases so numerous or important to justify the effort needed for reform? Why tamper with boundaries between the law's fiefdoms at the risk of undermining their stability and integrity?

I have no response to such questions that would convince the skeptical. Specifically defined results of the approach that I urge cannot be forecast. Nor can I predict improvement in the lot of the mentally ill or a clear gain in overall policy objectives. If the seriousness of present problems and the forceful criticisms of present practices and procedures given by the contributors to this volume and others are any indication, however, I should think the attempt at remedy worthwhile. If present demographic trends continue without catastrophic interruption, this attempt to work from broader analytical perspectives might well become obligatory – not because of any slight increases in the number of mentally ill persons as psychiatrists may define them now, but on account of substantial numbers of aged persons.

By the year 2000, it is projected that one in every five Americans will be 65 years of age or older. Mental illnesses, some of which stem from physical ailments that mimic the signs of dementia, are common disorders of the aging. Yet whether cause or symptom, these difficulties will occasion increasingly strident demands for attention to questions of competence and capacity across the law's broad spectrum. The management and testamentary disposition of property, consent to medical treatment, workmen's compensation and other insurance, welfare entitlement, differences between 'insanity' and 'senility', the legal significance to be placed on differences between chronic brain syndrome and hypochondriasis with chronic depression, assignment by relatives to nursing home 'care' without due process of law– these are only a beginning for a catalogue of the tough questions and issues that will accompany the graying of America.

It is therefore more in regard to the future than to the present that I feel urgent needs for wider ranging collaboration of philosophers and policy-makers. Constriction of our studies in law from a hardening of the categories cannot help us reach shared perceptions of what to do and how to go about it. Like the man on a train facing backwards, we shall see only that which is passed and passing, never that which is coming. In responding to the challenges of an aged and aging population, I am convinced we can afford neither the affliction nor the resulting blindness to impending change.

University of Texas Medical Branch
Galveston, Texas

NOTES

[1] I borrow the term from F. H. Underhill, see Toynbee ([6], p. 1).

[2] Stephen Wear [7] describes the figure of Thomas Szasz as one lurking behind and peering over the shoulders of this volume's contributors. I concur; the minority view which Szasz's arguments represent should encourage not merely direct refutation or defenses, but also study of real policy alternatives. Professor Creson's point, amplified by Wear, is particularly apropos in this connection: discarding the label and doing away with the metaphor of mental illness might well produce responses to socially deviant behavior more harmful and objectionable than those we now observe. Confident pronouncements as to the mythic character of mental illness may have somewhat the same measure of success achieved by papal bulls against comets: mental illness is banished from the language, yet Szasz's "problems in living" and the "bitter pill of moral conflicts in human relations" will remain.

This is not to say that the difficulties and dislocations of change can ever justify the status quo. I do suggest that policy options, collateral effects and special needs incident to social and cultural change are legitimate areas for inquiry along with purely conceptual matters. Determinations of what the problem is cannot be separated easily from decisions on what to do about it. Indeed, even if the philosophers should come to agree, physicians, legislators, judges, and their instruments will confront persons without a place in society who act at variance with its norms of behavior. Toward and against such persons there have always been responses of society backed, ultimately, by the physical power of the state. One expects there always will be. Real or imagined threats to property and safety, offenses to tastes and sensibilities of respectable folk – these have prompted action. We may be unsure about securing social stability through coercion, rehabilitation and welfare support of individuals, be they criminals, madmen or the 'undeserving' poor. That something will be done about these people, however, is not a question. It is a given.

BIBLIOGRAPHY

1. Coleman, J.L.: 1980, 'Mental Abnormality, Personal Responsibility and Tort Liability', this volume, pp. 107–133.
2. Creson, D. L: 1980, 'Function of Mental Health Codes in Relation to the Criminal Justice System', this volume, pp. 211–219.
3. Delkeskamp, C.: 1980, 'Criticial Use of Utilitarian Arguments: Szasz on Paternalism', this volume, pp. 177–207.

4. Moore, M. S.: 1980, 'Legal Conceptions of Mental Illness', this volume, pp. 25–69.
5. Neu, J.: 1980, 'Minds on Trial', this volume, pp. 73–105.
6. Toynbee, A.: 1961, *A Study of History: Reconsiderations,* Vol. XII, Oxford University Press, London, New York, Toronto.
7. Wear, S.: 1980, 'The Diminished Moral Status of the Mental Ill', this volume, pp. 221–230.

EDMUND L. ERDE

INVOLUNTARY CIVIL COMMITMENT: CONCERNING THE GROUNDS OF ETHICS

> The first man who went crazy was Cain, and like many of his successors, he had good reasons.
>
> *O. Friedrich* ([6], p. 40)

> Were you to preach, in most parts of the world, that political connexions are founded altogether on voluntary consent or a mutual promise, the magistrate would soon imprison you, as seditious, for loosening the ties of obedience; if your friends did not before shut you up as delirious, for advancing such absurdities.
>
> *David Hume* ([7], pp. 258–259)

> If the human is to be appropriately characterized as neither a biological system, nor a rational or meaning-producing entity, nor a collection or aggregate of fully formed individuals, then what is the appropriate view? Clearly, human beings *are* organisms, *are* thinking and reasoning beings, and *are* individuals or persons. The rub comes in abstracting such features from their very genesis in the concrete social and historical contexts in which the species first emerged, in evolutionary terms, and in which it developed its specific historical character, in postevolutionary or cultural terms .... This positive characterization of human ontology would then subordinate the biological, psychic, and personal or individual features to a more fundamental category: the socio-historical and cultural. What this comes to is a realization of individuality, of reason and of organic structure and function as the *products* of an historical development, and as essentially social both in their genesis and in their ongoing modes.
>
> *Marx Wartofsky* ([17], p. 72)

The essays contained in this book are about power. They are about the power of concepts, about the power of words, and the power of institutions, and the

B. A. Brody and H. Tristram Engelhardt, Jr. (eds.), Mental Illness: Law and Public Policy. 239–247.

power of persons. But the power in each case is not merely the power of a bulldozer, for a bulldozer's power is physical. The power of psychiatrists and judges is of a different kind, because they are empowered and empower. In other words, theirs is a social power. To a large extent, the issues which motivate the collection of the essays in this volume concern the proper scope and limits of social power. But social power is ethereal when it is possessed and exercised without physical powers. Thus, limitations on power are sometimes physical, if, say, the carburetor on a bulldozer is not working right, but sometimes questions of limitations arise about whether power *should* be compromised or contained – whether anyone could or should have the authority to order and use physical power to enforce anything as dubious as involuntary civil commitment, for example. Thus, beyond being about power, the essays in this book are also about social rationality. They concern themsevles with understanding rationality and its opposites (e.g., insanity) both in terms of what they mean and in terms of how such meanings do and should shape society.

One question which aligns these two foci (power and rationality) is how authorization arises, how moral power is established? Answering it may help us clarify the notion of involuntary civil commitment. In this essay, I shall argue for the following answer: If society is the ground of moral life as much as individuality is, then the *imposition* of rationality on and power over the prerational neophyte can help us understand and justify a form of involuntary civil commitment in spite of the good arguments against it.

Now, even within the contents of these essays there is a challenge. Professor Moore does not seem to think that the *issues* are very complex for he declares that "there is something about mental illness itself that excuses . . ." ([12], p. 39). He seems to want to use this axiomatically and stridently ([12], p. 62), although he does refine his position by referring to kinds and degrees of irrationality. There is the kind which applies to each of us to some degree, and there is the ancient paradigm of severely diminished rationality. Moore's analysis clarifies why individuals are exempt from culpability if they are irrational, but I want to consider why such persons should or should not be liable to subjection to civil commitment. For if mental illness is a prima facie excuse, why is it also not a prima facie justification of commitment? And, even if, as Moore suggests ([12], p. 36), the law dictates according to its own ends, why can those ends not be assessed rather than simply submitted to?

Dialectical reflections on policy often take the shape of a struggle between consequentialist-utilitarian thought on the one hand and deontological thought on the other. The issue involved in policies about commitment centers on the place of individual rights in contrast to social goods. It asks whether respect for personal freedom (deontological ethics) is morally inviolate, or whether consequences or accomplishments are more important than such respect (utilitarian ethics).

Considerations of irrationality and parentalism intrude into this tug in a crucial way since individuals or institutions may claim that their superior rationality licenses the imposition of power upon individuals or groups for the sake of the good consequences which they believe, given the circumstances, to be more commanding in spite of unpopularity or resistance. This behavior was decried by the classical utilitarian, John Stuart Mill, as Sartorius's essays shows, but the *logic* of parentalism is not blocked by the general theory of maximizing goods when what counts for goodness is different from what individuals desire, i.e., when the goodness is stipulated by society in, for example, social or religious terms. However, the position Mill and others hold is that freedom is a special value to be fostered, a special excellence to be enhanced. Such thinkers protect freedom for its utility, i.e., for the consequences freedom is thought to foster.

This position on freedom conflicts with the one offered by those who protect it due to transcendental or logical considerations. Those who found ethics with transcendental arguments start with at least these two premises: (1) our concept of responsibility is such that attributing it to an individual commits us to presupposing that he or she could have chosen otherwise and (2) the act one does is defined by the conception one has of what one is choosing to do. From here, transcendentalists argue that one must recognize that there is no such thing as ethics unless persons can freely choose what they wish to try to do. So, freedom to will, even more than freedom to act, is a necessary condition for the possibility of responsibility. When individuals are deprived of such freedom of thought, including the freedom to err or the freedom to willingly commit wrongs, their ethical and moral status is destroyed or denied.[1] This can be displayed by noting that although a world in which individuals could not think morally might be better in the sense of being more pleasant, the individuals in such a world would be no better in a moral sense than inanimate objects – no better than robots can be – even though they might be better off. In short, if the world were constituted by objects and robots, the entire moral universe would not exist. This argument for the logical priority of freedom of choice constitutes the meaning behind the logico-ethical maxim that 'ought' implies 'can'.

Having described the rival ethical postures on freedom, let us consider how each bears on involuntary civil commitment. There are some strong arguments of both the utilitarian and deontological variety which call such policies into question. The utilitarian's are well indicated by Sartorius. Although he is primarily concerned in the early part of his essay with paternalistic confinement – confinement for the sake of the confined – he rises to a more general level when, in agreeing with Mill, he argues that overall noninterference in the lives of others has greater utility than any alternative ([16], p. 142). Sartorius starts his essay by reminding us that (1) being "in need of treatment" and (2) being "dangerous (or potentially harmful) to oneself" are hopelessly vague ([16], p. 137). Given such vagueness we should try to define the clearest cases of individuals whom we would confine. These would be irrational agents who are therefore dangerous. A utilitarian might argue that confinement of such a person would produce greater good than harm. But, we would need a *perfectly accurate test* for finding *only* the dangerous. As Sartorius ([16], pp. 143–144) and Brock ([1], pp. 158–160) show, classical utilitarianism suggests that, because the false positive rate of even a very good test would involve a great number of mistaken confinements, we should not allow for such confinement. The classical consequentialist seems thereby committed to the unacceptability of involuntary civil commitment policies, especially because diagnoses in psychiatry are notoriously unreliable [15]. And Moore's distinction (i.e., between the kinds of irrationality which infect each of us and the kind which is mental illness) will not approach felicitous applicability to cases.

So whether one is being paternalistic – confining for the sake of the confined – or trying to protect society at large, one cannot accept the proportion of errors inevitably involved in sorting the putatively well from the putatively sick.

Deontologically,[2] as Brock argues it, paternalistic intervention against other persons can never be justified. The *rational* independence of others must always be respected even when they are adjudged in error ([1], pp. 160ff). Thus, we must consider how to construe irrationality (when this is taken as equivalent to mental illness).

If persons are mentally ill in a severe and persistent way and not merely irrational in the typically human way, they cannot, according to Moore, frame practical syllogisms ([12], p. 62). Can they then feel imposed upon by anything? It would seem not. This is probably too extreme or demanding a view of what is required for an excuse on grounds of mental illness. It also may be too extreme in that we would not be able to institutionalize those whom we feel ought clearly to be institutionalized against their will.

The grounds permitting the incarceration of individuals should vary depending on whether they are persons (rational), or whether they can only be conceived as human robots. For example, to take a deontological stance, anyone who wills against another's freedom is willing implicitly to lose his or her own freedom and can therefore be punished, confined, etc., as a person.[3] This makes it look as though we can hold responsible the rational violators of the moral order and just hold onto the human robots, the haywired persons (former persons), who, in Coleman's discussion, cannot even understand the reprisals that may be taken against them for their actions ([3], pp. 116–117).

It looks as though we can, then, legitimately punish (perhaps through confinement) those rational agents who violate those moral norms which have a prescribed legal sanction and it looks as though we may rightfully confine those troublesome beings who are not persons because they are not rational.

These reflections on giving reasons and controlling irrational entities suggest that the concept of *rationality* is a bridge between philosophy and psychiatric research. One definition of rationality is implied by Neu: "The *capacity* to correct [adjust] understanding, to respond to his relevant reality" ([13], p. 86, cf. p. 89) which coheres with Moore's ideas on practical syllogisms; also cf. [5]). However, a jarring blow follows once we begin to see how socially informed our ideas of diseases (cf. [11, 10]) and rationality are. Putting that together with aspects of a concept of society such as the one Creson contributes, we must feel stunned. No answer about a good, fair, easily applicable policy of civil commitment is likely ever to be forthcoming.

Rationality, language and society are integrally related. Consider just Neu's notion as an example. The concepts *relevant* and *reality* can only be defined and tested through the public language. Defining and testing what we mean by *adjustment of understanding* has the same dependence on language and its intrinsically social, public nature [9]. The support and direction of a community is a necessary condition for learning a first language. The elders of the community teach the neophyte. For example, they teach that there is a distinction between being correct and being incorrect and they provide paradigm cases of these. They provide their members with rationality, i.e., both a catalogue of reasons for certain claims and truths and a method for reassessing, refining, and revising that catalogue. This suggests a model of irrationality and the model is borne out by Roger Brown [2] who (as a learning theorist) may be an expert and objective observer of mental illness. Brown reports his conclusions of a study of schizophrenia in the following vague but useful terms. Such persons' reality testing (admitting cultural variations in reality) is bad; they take their delusions seriously enough to base actions upon them, but are insufficiently sensitive to the negative consequences their actions may produce (cf. [3]).

This model of irrationality seems to hold not only in terms of the way persons perceive their physical world, the world which the natural sciences address, but it also seems to hold in terms of how persons understand their moral existence. In fact, it may be more elegant[4] to think of rationality as possessing two aspects. In this way, irrationality may be thought to involve both a distortion of perception and a set of deviant behaviors – behaviors which cannot be given a rational account. Together these aspects indicate a deviance from common *evaluations* and thus a failure to participate in the rational order which defines the entire fabric of community life, the community's paradigm (to make unauthorized extension of Kuhn's [8] sense).

Claiming that rationality is a unity of thought and behavior, and marking the relevance of both perception and evaluation, may help bridge the nagging schism between facts and values and explain why irrationality seems to be threatening in that irrational persons do not seem to understand the way the world works; they do not seem concerned to practice care on behalf of those objects which are esteemed, and they cannot be well controlled by the deterrents involved in punishment or self-destructive activity.

Irrationality may then be a vitally important subset of those "failures to learn what society tries to teach". This leads me to a brief consideration of the implications of Dr. Creson's contribution to this volume and its import for ethics and civil commitment. The force and bearing of Creson's ideas are especially enlarged by what Professor Margolis has to say about both prudence and about persons as culturally emerging entitles. For not only is free will a *necessary condition for the possibility* of moral existence but, as Creson, Margolis and many others argue, social rearing is a *necessary condition of* human rationality.

This is to say that society should be construed as having an existence and nature greater than the mere collection of individuals (as Plato's *Crito* is meant to show). Creson's essay can be read as a description of that nature, or as a warning about the limitations or tolerances or homeostatic reactions of society because he suggests that certain ideologic commitments may be causally impossible to fulfill. Living up to some commitments might cause a breakdown in other functions which are embodied in policies based on them or on some other ideals. Specifically, the civil commitment policies of "mental health codes may be viewed as an escape valve that makes possible a society's self-righteous, albeit hypocritical regard for individual rights" ([4], p. 218).

Creson offers two alternative consequences regarding commitment: (1) to the extent one relinquishes involuntary civil commitment policies, one may to that degree force a reconceptualization in terms of a criminalization of actions now not considered crimes *or* (2) to the extent one medicalizes behavior now thought to be criminal, one may lose the moral authorization for certain kinds of control over problematic agents.

Both the social features of rationality, and the fact that there is a spectrum of rationality (or mental skill) are vitally important in clarifying the nature and justification of involuntary civil commitment. Through imposing a language on persons, society imposes a standard of rationality on each of us – which is meant to include the realm of facts and the realms of values and morality. Society and language therefore impose a way of life. All this implies that the basis for the existence and implementation of a policy of involuntary civil commitment may be wider than just the presence of fear, anxiety or disgust caused by aberrant behavior. These emotions may exist (one may speculate) because the social paradigm, the form of life, the ground of human existence, is apparently threatened by certain problematic behavior.

So perhaps Creson is supercritical when he tells us we may be hypocritical in having policies and practices that permit involuntary civil commitment. Perhaps, too, this is the force of the insight shared by radical psychiatrists such as Laing and Szasz – i.e., that we can appreciate why on occasion persons resort to madness, why, as O. Friedrich put it, Cain may have had good reasons for going crasy – the experiences of such persons make them need, and sometimes seek, help and care.

But because going mad affronts our form of life, we might retain the policy of involuntary civil commitment despite the excellent deontological and consequentialist arguments that speak against *practicing* that policy. If we heed these sound arguments, recall Creson's prognostication, and maintain the concern expressed in the essays of this volume with respect to a democratic, libertarian form of life, we are compelled to conclude that we should *keep the policy and use it as little as possible.* To do otherwise might be imprudent, and, as Joseph Margolis remarked elsewhere, "Prudential interests ... are enabling interests, that is, the general (determinable) condition on which any ethical, political, economic program viable for a complex society must depend; in that sense, the pursuit of prudential interests is prima facie rational ..." ([10], p. 252). Here Margolis's point seems to be identical with the one made by Wartofsky in the third epigram preceding this essay.

University of Texas Medical Branch
Galveston, Texas

NOTES

[1] By using the disjunction 'destroyed or denied' I am trying to mark aspects of the impact of inconsistency. To err in arithmetic is different from trying to get away with (say) falsifying figures. Juggling the books appeals to the foundational features of number theory in ways which would *destroy* arithmetic were those appeals universally accepted.

[2] Under a deontological ethic, no one can be used as a mere means to the ends of others. Deontologists hold that acts directed against other rational agents must respect their freedom to act or be in violation of the logical requirements of morality. Whether one is a deontologist or consequentialist, however, one's position is incomplete without the following logical principle of ethics: one must be willing to universalize the warrant for an action, be willing to have it used to justify an action taken against the interest of any individual including oneself. Thus, for example, if we deprive individuals of information or a course of action which we (only) think they should not want to know or have open to them, we are in effect depriving all, even ourselves, of truths and freedoms. The moral bearing of such conduct is that it destroys or denies the moral integrity of the person's action because it is a self-denial of the practice or claim to practice his or her ability to ever tell the truth or to act freely because lying and coercion would become universal. Logic, therefore, requires that the truth be told and that freedom and rights of others be respected. In this way ethics can be conceived as a search for rules to which individuals ought to be willing to commit themselves before they attend to the particulars of the lives they would lead. This is the power of Rawls's [14] 'original positon'.

To be a deontologist is, then, to be one who on transcendental grounds holds autonomy to be inviolate and subject to universalization. When this principle is endorsed, involuntary civil commitment becomes extremely difficult to sustain.

[3] Anyone who wills in this way also demonstrates an implicit willingness to violate moral logic and is therefore morally irrational. This is different from just going mad. We can, as deontologists, take one's unjustifiable violation of a moral rule as a statement of one's moral willingness to have one's own freedom compromised, and we can take that as a moral authorization of punishment. Alternatively, we can take a willingness to abandon moral logic as the display of moral irrationality which would warrant controlling that morally irrational entity. Such control is warranted because the individual may still be able to act in such away as to have power enough to do harm.

[4] Moore tells us, in the passages where he discusses the use of terms like 'mental illness', that these terms are descriptive of behavior rather than explanatory by way of causal connection. Neu, however, suggests that the law coheres with (causes?) our intuitions, that there are such things as mental capacities and conditions which, when they go awry, may influence the responsibility-status to which individuals are assigned. In the present

discussion I believe that I am synthesizing the factual (perceptual) distortion and the value (evaluative) deficiency that are appealed to when we judge individuals mentally ill. This suggests that, although the two ways of characterizing the use of terms connected with mental illness (which Moore and Neu discuss) are thoroughly different, we may still mark each as correct. This also indicates that we must, even in theorizing about such important matters, come to recognize our own irrationality. In fact, perhaps the most significant message of this essay is precisely that. As the later discussion attempts to show, we seem compelled to accept practice we cannot really justify. In the end we are at best forced to appeal to either prudence, hypocrisy, or mystery.

BIBLIOGRAPHY

1. Brock, D. W.: 1980, 'Involuntary Civil Commitment: The Moral Issues', this volume, pp. 147–173.
2. Brown, R.: 1973, 'Schizophrenia, Language, and Reality', *American Psychologist* 28, 395–403.
3. Coleman, J.L.: 1980, 'Mental Abnormality, Personal Responsibility and Tort Liability', this volume, pp. 107–133.
4. Creson, D.L.: 1980, 'Function of Mental Health Codes in Relation to the Criminal Justice System', this volume, pp. 211–219.
5. Erde, E.L.: 1977, 'Round Table Discussion', in H. T. Engelhardt, Jr. and S. F. Spicker (eds.), *Mental Health: Philosophical Perspectives,* D. Reidel Publ. Co., Dordrecht, Holland, pp. 285–293.
6. Friedrich, O.: 1975, *Going Crazy,* Simon and Schuster, New York.
7. Hume, D.: 1965, 'Of the Original Contract', in A. C. MacIntyre (ed.), *Hume's Ethical Writings,* Collier Books, New York.
8. Kuhn, T.: 1970, *The Structures of Scientific Revolutions,* 2nd ed., University of Chicago Press, Chicago, Ill.
9. Malcolm, N.: 1966, 'Wittgenstein's Philosophical Investigations', in George Pitcher (ed.), *Wittgenstein,* Doubleday and Co., Garden City, New York, pp. 65–103.
10. Margolis, J.: 1976, 'The Concept of Disease', *The Journal of Medicine and Philosophy* 1, 252–3.
11. Margolis, J.: 1980, 'The Concept of Mental Illness: A Philosophical Examination', this volume, pp. 3–23.
12. Moore, M. S.: 1980, 'Legal Conceptions of Mental Illness', this volume, pp. 25–69.
13. Neu, J.: 1980, 'Minds on Trial', this volume, pp. 73–105.
14. Rawls, J.: 1971, *A Theory of Justice,* Belknap Press of Harvard University Press, Cambridge, Massachusetts.
15. Rosenhan, D. L.: 1973, 'On Being Sane in Insane Places', *Science* 179, 250–258.
16. Sartorius, R. E.: 1980, 'Paternalistic Grounds for Involuntary Civil Commitment: A Utilitarian Perspective', this volume, pp. 137–145.
17. Wartofsky, M. W.: 1975, 'Organs, Organisms and Disease: Human Ontology and Medical Practice', in H. T. Engelhardt, Jr. and S. F. Spicker (eds.), *Evaluations and Explanation in the Biomedical Sciences*, D. Reidel Publ. Co., Dordrecht, Holland, pp. 67–83.

NOTES ON CONTRIBUTERS

Dan W. Brock, Ph.D., is Associate Professor of Philosophy, Brown University, Providence, Rhode Island.

Baruch A. Brody, Ph.D., is Professor of Philosophy and Chairman, Department of Philosophy, and Director of Legal Studies Program, Rice University, Houston, Texas.

Jules Coleman, Ph.D., is Associate Professor of Philosophy, University of Wisconsin, Milwaukee, Wisconsin.

Daniel L. Creson, M.D., Ph.D., is Clinical Associate Professor of Psychiatry, University of Texas Medical Branch, Galveston, Texas, and Clinical Associate Professor of Psychiatry, University of Texas Medical School, Houston, Texas.

Corinna Delkeskamp, Ph.D., is Adjunct Assistant Professor of Philosophy, Pennsylvania State University, University Park, Pennsylvania.

H. Tristram Engelhardt, Jr., Ph.D., M.D., is Rosemary Kennedy Professor of the Philosophy of Medicine, Kennedy Institute of Ethics, Georgetown University, Washington, D.C.

Edmund L. Erde, Ph.D., is Associate Professor of the Philosophy of Medicine, Institute for the Medical Humanities and the Department of Preventive Medicine and Community Health, University of Texas Medical Branch, Galveston, Texas.

Joseph Margolis, Ph.D., is Professor of Philosophy, Temple University, Philadelphia, Pennsylvania.

Michael S. Moore, J.D., is Associate Professor of Law, University of Southern California, Los Angeles, California.

Jerome Neu, Ph.D., is Professor of Philosophy, University of California at Santa Cruz, Santa Cruz, California.

Rolf Sartorius, Ph.D., is Professor of Philosophy, University of Minnesota, Minneapolis, Minnesota.

James B. Speer, Ph.D., J. D., is Assistant Professor of the History of Legal Medicine, Institute for the Medical Humanities, University of Texas Medical Branch, Galveston, Texas.

Stephen Wear, Ph.D., is Assistant Professor of Philosophy and Coordinator of the Program in Humanities and Medical Ethics, State University of New York School of Medicine, Buffalo, New York.

INDEX

The Philosophy and Medicine Book Series

Editors

Stuart F. Spicker and H. Tristram Engelhardt, Jr.

1. **Evaluation and Explanation in the Biomedical Sciences**
   Proceedings of the First Trans-Disciplinary Symposium on Philosophy and Medicine, Galveston, Texas, May 9–11, 1974
   1975, vi + 240 pp. ISBN 90-277-0553-4

2. **Philosophical Dimensions of the Neuro-Medical Sciences**
   Proceedings of the Second Trans-Disciplinary Symposium on Philosophy and Medicine, Farmington, Conn., May 15–17, 1975
   1976, vi + 274 pp. ISBN 90-277-0672-7

3. **Philosophical Medical Ethics: Its Nature and Significance**
   Proceedings of the Third Trans-Disciplinary Symposium on Philosophy and Medicine, Farmington, Conn., Dec. 11–13, 1975
   1977, vi + 252 pp. ISBN 90-277-0772-3

4. **Mental Health: Philosophical Perspectives**
   Proceedings of the Fourth Trans-Disciplinary Symposium on Philosophy and Medicine, Galveston, Texas, May 16–18, 1976
   1978, xxii + 302 pp. ISBN 90-277-0828-2

5. **Mental Illness: Law and Public Policy**
   edited by Baruch A. Brody and H. Tristram Engelhardt, Jr.
   1980, xvii + 254 pp. ISBN 90-277-1057-0

6. **Clinical Judgment: A Critical Appraisal**
   Proceedings of the Fifth Trans-Disciplinary Symposium on Philosophy and Medicine, Held at Los Angeles, California, April 14–16, 1977
   edited by H. Tristram Engelhardt, Jr., Stuart F. Spicker, and Bernard Towers
   1979, xxvi + 278 pp. ISBN 90-277-0952-1

7. **Organism, Medicine, and Metaphysics**
   Essays in Honor of Hans Jonas on his 75th Birthday, May 10, 1978
   1978, xxvii + 330 pp. ISBN 90-277-0823-1

GPSR Compliance

The European Union's (EU) General Product Safety Regulation (GPSR) is a set of rules that requires consumer products to be safe and our obligations to ensure this.

If you have any concerns about our products, you can contact us on ProductSafety@springernature.com

In case Publisher is established outside the EU, the EU authorized representative is:

Springer Nature Customer Service Center GmbH
Europaplatz 3
69115 Heidelberg, Germany

Batch number: 10823811

Printed by Printforce, the Netherlands